SpringerBriefs in Stem Cells

For further volumes:
http://www.springer.com/series/10206

Tiziana A.L. Brevini • Fulvio Gandolfi

Pluripotency in Domestic Animal Cells

 Springer

Tiziana A.L. Brevini
Laboratory of Biomedical Embryology
  Anatomy and Histology
Università degli Studi di Milano Unistem
Milan, Italy

Fulvio Gandolfi
Laboratory of Biomedical Embryology
  Anatomy and Histology
Università degli Studi di Milano Unistem
Milan, Italy

ISSN 2192-8118   ISSN 2192-8126 (electronic)
ISBN 978-1-4899-8052-6   ISBN 978-1-4899-8053-3 (eBook)
DOI 10.1007/978-1-4899-8053-3
Springer New York Heidelberg Dordrecht London

Library of Congress Control Number: 2013954542

Springer is part of Springer Science+Business Media (www.springer.com)

Preface

The establishment of stem cell (SC) lines in large animal species could lead to important improvements in the biomedical field as well as in agriculture and animal science.

In particular, domestic animal SC would have a positive impact on the health and production traits of these species through genetic engineering. At the same time, because of their great similarity to humans in morphological, physiological, and immunological functions, large animals are a very effective and suitable animal model for biomedical studies and preclinical trials, allowing us to select the most appropriate species for a specific pathology to be addressed.

However, although proven mouse and human SC lines have been established, no fully validated SC lines are available in large animal species. We discuss here some of the factors that make the establishment of SC lines in animal species other than mouse and human a very slow process and address aspects related to the specific mechanisms ensuring and controlling pluripotency and cell commitment in these species. We believe that this topic is meeting increasing interest because of the great potential of domestic animal species as biomedical models intermediate between the mouse and the human.

Data from the literature suggest that similar regulatory pathways are likely to exist among different species. Coupling of these pathways with their distinct expression patterns, the relative concentrations of pluripotency-related molecules, the timing of embryo development, and specific microenvironmental conditions all vary in a species-specific manner. We believe that understanding of these subtle but meaningful diversities may provide beneficial information about the isolation of genuine stem cells in large animals.

This brief is intended as a handy, easy-to-read overview of the main concepts related to stem cells in large animal species with their potential applications in genetic engineering and animal modeling for preclinical biomedical applications.

Milan, Italy
Tiziana A.L. Brevini
Fulvio Gandolfi

Contents

1 Early Embryo Development in Large Animals 1
   1.1   Syngamy and Spindle Formation 1
   1.2   Cleavage, Compaction, and Blastulation 6
   1.3   Cell Commitment 8
   1.4   Naïve Versus Established Epiblasts 12
   Further Reading 18

2 Pluripotency in Domestic Animal Embryos 21
   2.1   Expression, Restriction, and Interactions of Molecules
         Involved in the Control of Cell Potency and Commitment 21
   2.2   Species-Specific and Stage-Specific Pluripotency Markers 23
   Further Reading 26

3 Use of Large Animal Models for Regenerative Medicine 29
   3.1   Why Do We Need Stem Cell Lines from Large Animal Species? 29
   3.2   What Are the Applications and Potentials? 31
   3.3   A Lesson from the Heart 32
   3.4   Therapies with Mesenchymal Stem
         Cells in the Horse… Are We There? 35
   3.5   What Are the Limits? 39
   Further Reading 41

Acknowledgments 43

About the Authors 45

Chapter 1
Early Embryo Development in Large Animals

Abstract In this chapter, all the main events taking place during the first phases of embryo development are examined. Syngamy and the first mitotic spindle formation are described, as well as the processes of cleavage, compaction, and blastulation. The importance of the integrity of the first mitotic spindle is highlighted because it is fundamental for normal and correct embryo development and cell commitment. Gastrulation and subsequent inner cell mass cell differentiation with the definition of the three germ layers and establishment of the body axis are also illustrated. In fact, each cell constituting an embryo is the target, as well as the source, of specific signals. These signals regulate cell differential gene expression and epigenetic restrictions that gradually limit cell potency to a phenotype-related expression pattern. The hypothesis according to which each cell present in the embryo maintains a memory of its own proliferation history and positional changes during development is also discussed.

1.1 Syngamy and Spindle Formation

The process by which the spermatozoon and the egg unite is known as fertilization. In mammalian species, this event occurs in the ampullary region of the oviduct. In domestic animals, with the exception of dogs, the oocyte arrested at metaphase of meiosis II enters the oviduct via the infundibulum. It is approached by the capacitated spermatozoa, which interact and penetrate the zona pellucida (ZP) and then fuse with the oocyte plasma membrane. Immediately after the entry of the spermatozoon, the oocyte undergoes the activation process that induces the cortical reaction, stimulates the second polar body extrusion, and leads to initiation of embryonic development. In particular, at the time of fertilization the female gamete contains, in its pro-nucleus, a haploid set of chromosomes, as well as its male counterpart. The latter increases its volume by chromatin decondensation, after the entry into the oocyte, through a reduction of disulfite crosslinks, made by glutathione, and a

T.A.L. Brevini and F. Gandolfi, *Pluripotency in Domestic Animal Cells*, SpringerBriefs in Stem Cells, DOI 10.1007/978-1-4899-8053-3_1, © The Author(s) 2013

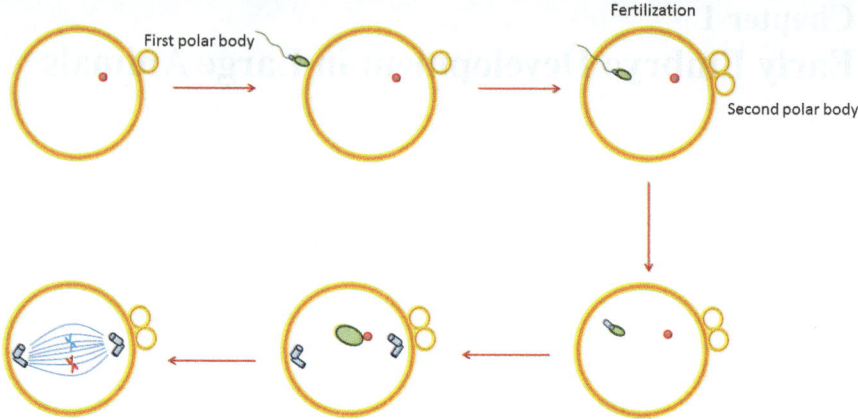

Fig. 1.1 Fertilization and syngamy. At fertilization, the arrested MII oocyte extrudes the second polar body. The male pro-nucleus increases its volume by means of chromatin decondensation and approaches the female pro-nucleus, moving toward the center of the oocyte. At syngamy, male and female pro-nuclei are fused, giving rise to a diploid genome distinctive of the zygote

replacement of sperm polyamines with histones from the oocyte. Finally, the sperm centrosome organizes the aster that captures the female pro-nucleus, drawing it near the male pro-nucleus, and these move toward each other in the center of the oocyte. At the time of nuclear envelope dismantling, the pro-nuclei are fused and syngamy takes place. The diploid genome of the zygote is thus created (Fig. 1.1).

One of the first fundamental events necessary for correct early embryo cleavage is the formation of a new mitotic spindle. The spindle comprises spindle microtubules that originate at the centrosome, a non-membrane-bound cytoplasmic organelle composed of several protein complexes that organize interphase microtubule arrays (Fig. 1.2).

In this context, it is important to elucidate the essential role played by the sperm centrosome during fertilization; this is possible by taking a step back and carefully considering the gamete maturation process.

Initially spermatids and primary oocytes show, in common with somatic cells, a typical centrosome organization, characterized by the presence of a pair of centrioles surrounded by pericentriolar material.

This somatic cell-like centrosome undergoes extensive modification or degeneration during the final stages of gametogenesis to meet the specific needs of gamete function and fertilization.

Indeed, in mammals, centrosomes are reduced during spermiogenesis, and mature spermatozoa retain the centrioles but lose most of the pericentriolar centrosomal proteins. More in detail, mature male gametes of large animals show an intact proximal centriole (Fig. 1.3) whereas the distal centrioles are mostly disorganized or highly degenerated together with γ-tubulin and centrosomal proteins.

In contrast, during oogenesis, female gametes lose the centrioles and retain only a stockpile of centrosomal proteins. Until the pachytene stage, in fact, fetal oocytes

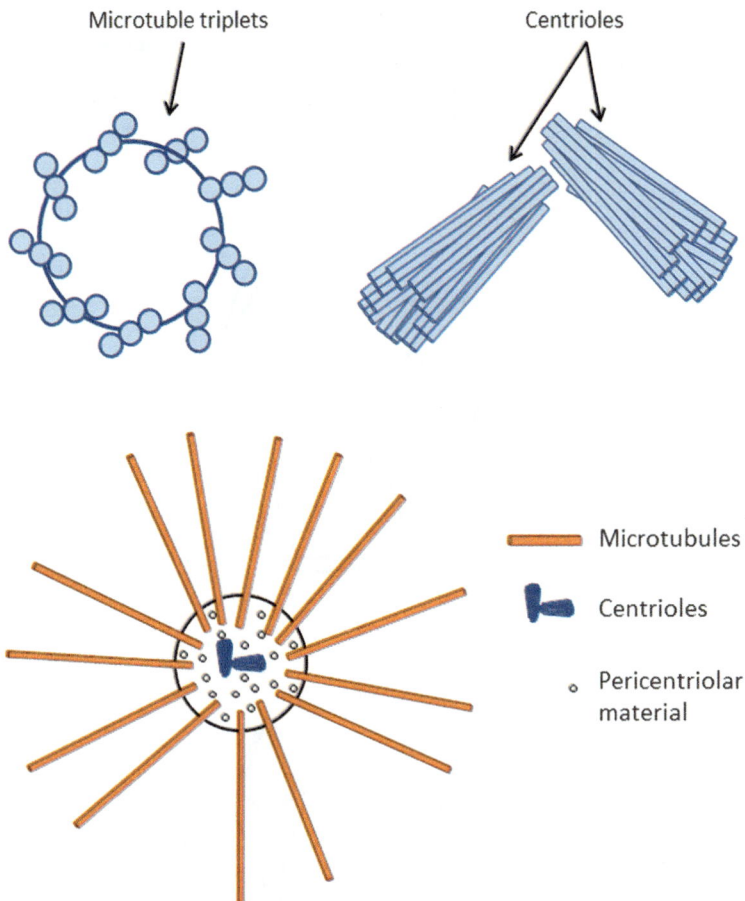

Fig. 1.2 Centrosome structure. Centrosomes are composed of two orthogonally arranged cylinder-shaped centrioles (*blue*) surrounded by an amorphous matrix of electron-dense proteins referred to as pericentriolar material (PCM; *white circle*). Centrioles contain centrin, cenexin, and tektin and display a 9+0 pattern of nine triplet microtubules and no central pair of microtubules. The PCM contains a protein complex responsible for microtubule (*orange*) nucleation and anchoring, including γ-tubulin, pericentrin, and ninein

and oogonia display a normal set of centrioles although these organelles are absent in the mature oocytes (Fig. 1.4). This degenerative process has been demonstrated in many large animals, including sheep, cows, and pigs.

Because of this reciprocal reduction of centrosomal constituents, sperm and mature oocytes are complementary to each other. Indeed, in large animals, early embryo development requires maternal and paternal contribution and, in particular, needs their elements to restore a normal and functional centrosomal structure. In fact, although in rodents there is no evidence of a functional centriole in the sperm

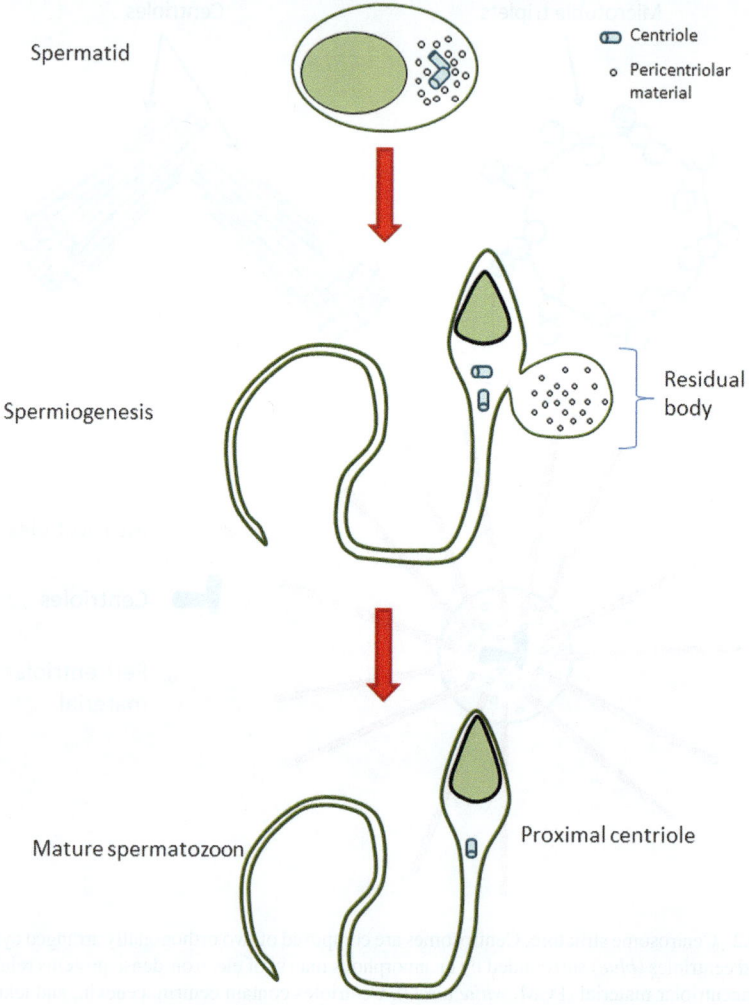

Fig. 1.3 Centrosome reduction in male gamete. Spermatids possess an intact centrosome containing centrioles and centrosomal proteins. During spermiogenesis, centrosomes are reduced, and the mature spermatozoon retains only the intact proximal centriole

and they are maternally inherited, in other mammalian species, the spermatozoa contribute the proximal centriole during fertilization (Fig. 1.5).

At the time of fertilization, sperm and egg equally contribute haploid genomes as well as the relative centrosome components, forming a new functional centrosome in the zygote. In agreement with this, the sperm tail and its centriole-harboring connecting piece are incorporated into the ooplasm together with the sperm head whereas most of the other sperm cytoplasmic structures including mitochondria, microtubules, and fibers are eliminated. The sperm head decondenses and the

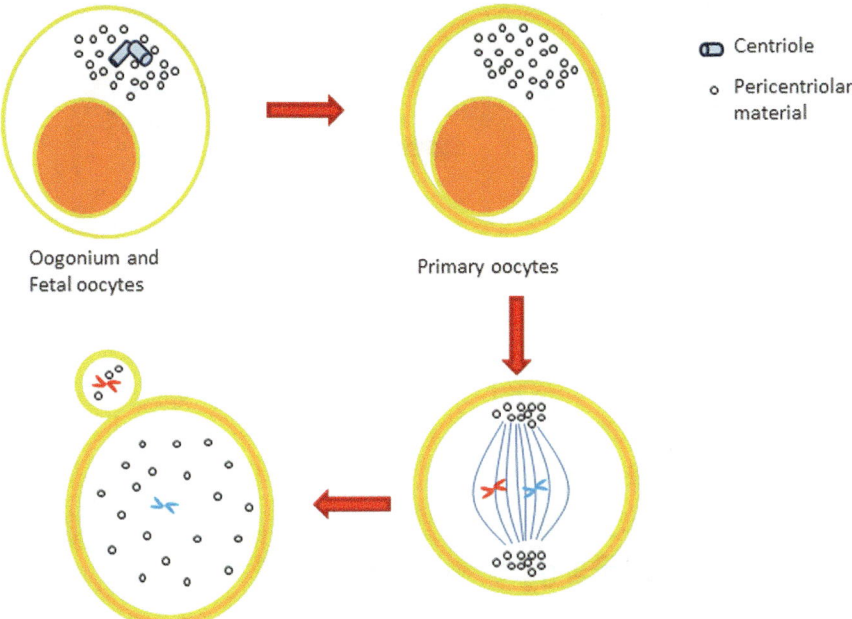

Fig. 1.4 Centrosome reduction female gamete. Oogonia possess standard centrosomes containing centrioles and centrosomal proteins. During oogenesis during spermiogenesis, centrosomes are reduced and mammalian primary oocytes display a complete loss of both centrioles, resulting in acentriolar and anastral poles during meiotic I and II divisions. During the nondividing stages, the pericentriolar proteins are dispersed in the cytoplasm of the oocyte, whereas during dividing stages, they are distributed as concentric poles of the barrel-shaped spindles

proximal centriole remains intact, forming the sperm aster sited around the male pro-nucleus subjacent to the oocyte cortex. The sperm aster enlarges and moves within the oocyte cytoplasm, ensuring male and female pro-nuclei apposition and formation of a single mitotic metaphase plate with a bipolar spindle.

In agreement with this, several studies, carried out in cattle and sheep, have demonstrated that, following insemination, sperm centrioles duplicate during the pronuclear stage, and at syngamy, centrioles are located at opposite poles of the first mitotic spindle (first embryo cleavage), together with the surrounding pericentriolar material of the oocyte that nucleates the microtubules.

Together, these observations suggest that zygote centrosomes represent the origin of embryonic, fetal, and adult somatic cell centrosomes.

In contrast, a recent study by Manandhar et al. revealed that in the pig, sperm centrioles are lost in the zygote after in vitro fertilization and are not detectable until the late blastocyst stage. Accordingly, early preimplantation cleavages show broad and anastral spindle poles, whereas only blastomeres of the hatched blastocysts develop centrioles comparable with those of culture cells.

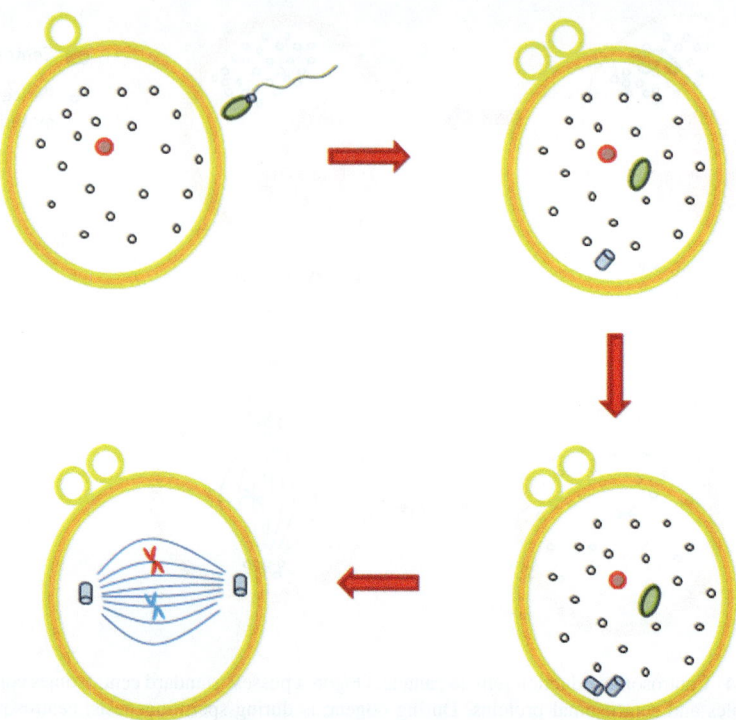

Fig. 1.5 Centrosome inheritance in domestic animals. The mature male gamete contains a proximal centriole, whereas the MII oocyte displays a meiotic spindle with acentriolar centrosome. At the time of fertilization, the proximal centriole is introduced by the sperm in the oocyte cytoplasm. It replicates forming the aster that, at syngamy, relocates to opposite poles to form the centers of the mitotic spindle poles and drives the first embryo mitotic division

1.2   Cleavage, Compaction, and Blastulation

At the time of fertilization, meiosis is completed and cells return to the mitotic cell cycle. The zygote has inherited from the oocyte the complete molecular and structural composition necessary to initiate the first cleavage: this is usually completed within 24 h (depending on the species; see Table 1.1) after ovulation and leads to the formation of two blastomeres that contain a full copy of the new embryo genome. A series of successive mitotic divisions take place without cellular growth and the cells become smaller and smaller. During this development the embryo maintains the same total volume because the original cytoplasm is split among the newly formed blastomeres and remains surrounded by the zona pellucida for several days.

Embryonic cleavage begins during the transport of the embryo along the maternal oviduct. The embryo then enters the uterus to implant. This event is characterized by species-specific timing (Table 1.2).

During oocyte maturation, maternal transcripts and proteins are stored in the ooplasm. These molecules are required for driving initial embryonic development

Table 1.1 Timing of preattachment embryogenesis in domestic species and in the human

Species	Two-cell stage	Four-cell stage	Eight-cell stage	Morula	Blastocysts	Hatching	Gestation
Bitch (dog)	3–7 days	–	–	–	–	13–15 days	60 days
Cow	24 h	1.5 days	3 days	4–7 days	4–10 days	9–11 days	270 days
Ewe	24 h	1.3 days	2.5 days	3–4 days	4–10 days	7–8 days	150 days
Mare	24 h	1.5 days	3 days	4–5 days	6–8 days	7–8 days	330 days
Sow	14–16 h	1 day	2 days	3.5 days	4–5 days	6 days	114 days
Woman	24 h	2 days	3 days	4 days	5 days	5–6 days	270 days

Table 1.2 Timing and related embryo stages of passage from the oviduct into the uterus

Species	Stage	Days after ovulation
Pig	4–8 cells	2
Sheep	8–16 cells	3
Cow	8–16 cells	3–3.5
Horse	Morula	5–6
Dog	Blastocyst	8

Table 1.3 Timing of genome activation in different species

Species	Genome activation
Pig	Four-cell stage
Cow	Eight-cell stage
Dog	Eight-cell stage
Horse	Eight- to 16-cell stage
Sheep	Eight- to 16-cell stage

and govern at least the first cleavage. After fertilization, in fact, transcripts and proteins present in the oocyte are gradually degraded and, at a species-specific stage of development, embryonic genome activation is required (Table 1.3). The initial transcription is very limited and increases later.

The following phase is characterized by the formation of the morula, formed by a large number of identical blastomeres. Subsequently, the blastomeres start to change and differentiate, under the control of specific transcription factors, namely CDX2 and EOMES (Fig. 1.6). This event is known as compaction and gives the embryo a smoother surface. In particular, the outer cells constitute the trophectoderm, or trophoblast cells, that attach with neighboring cells and form tight junctions and desmosomes. These specialized intercellular structures contribute to intercellular sealing and tissue integrity, critical for vectorial transport and blastocoele cavity formation.

Moreover, compaction is also characterized by species-specific timing: in the pig it occurs at the 8-cell stage (early embryo development), whereas in cattle it happens around the 16- or 32-cell stage (later embryo development).

The following step is known as the blastulation phase, during which the trophoblast cells secrete a fluid into the central cavity, the blastocyst cavity or blastocoele, lining the cavity. This event transforms the embryo into a blastocyst and usually

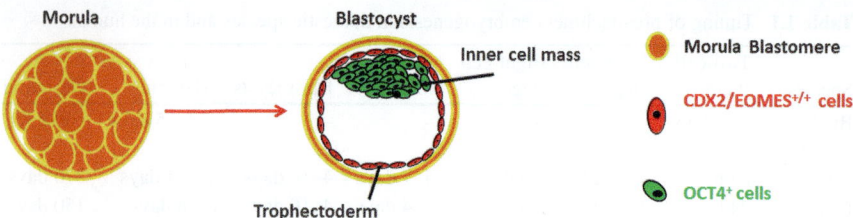

Fig. 1.6 Representative scheme of first differentiation process. After morula compaction, blasto-meres begin to change and differentiate. Some cells activate the expression of CDX2 and EOMES genes and others continue to express OCT4. The first group of cells gives rise to the trophectoderm cells; the second group maintains the pluripotent status, forming the inner cell mass cells

occurs during the first week of development. One pole of the blastocyst is occupied by the inner blastomeres, which give rise to the inner cell mass (ICM) and will form the embryo proper. In contrast, the other cells form the trophectoderm and will give rise to the embryonic placenta. The activity of a sodium pump located in the cell membranes of the trophoblast cells causes a blastocyst volume increase, drawing water into the central cavity. The blastocyst expansion may lead to rupture of the zona pellucida, which allows the blastocyst to escape through the opening. When this phenomenon does not occur, a blastocyst-secreted protease, known as strypsin, and proteolitic enzymes produced by the endometrium degrade and lyse the glyco-proteins forming the zona pellucida. This process, known as hatching, enables the trophoblast cells to directly bind to the uterine cavity (Fig. 1.7).

Around the time of hatching, ICM cells further differentiate into two cell popula-tions, the hypoblast and the epiblast cells (Fig. 1.8), and the blastocyst cavity becomes flattened and delaminates. To obtain the first type of cells, the hypoblast cells, activation of GATA-binding factor 6 (GATA-6), a fundamental transcription factor driving the formation of the primitive endoderm, is necessary. These cells have an epithelial morphology, with cuboidal shape closer to the blastocoele that will form the inner epithelium of the yolk sac. The second type of cells, the epiblast cells, will form the embryo proper and all the different tissue types that can be found in an organism. Furthermore, epiblast cells need the expression of NANOG, and their proliferation and differentiation are regulated by the hypoblast through forma-tion of a basement membrane.

1.3   Cell Commitment

During the embryo development, cells derived from one single zygote are able to differentiate, taking on many fates in the body, including all of the more than 230 different cell types present in the adult mammalian organism. These embryo-derived cells respond to specific stimuli at a specific time and, at the end of the differentia-tion process, each cell is highly committed to a distinct determined fate, giving rise

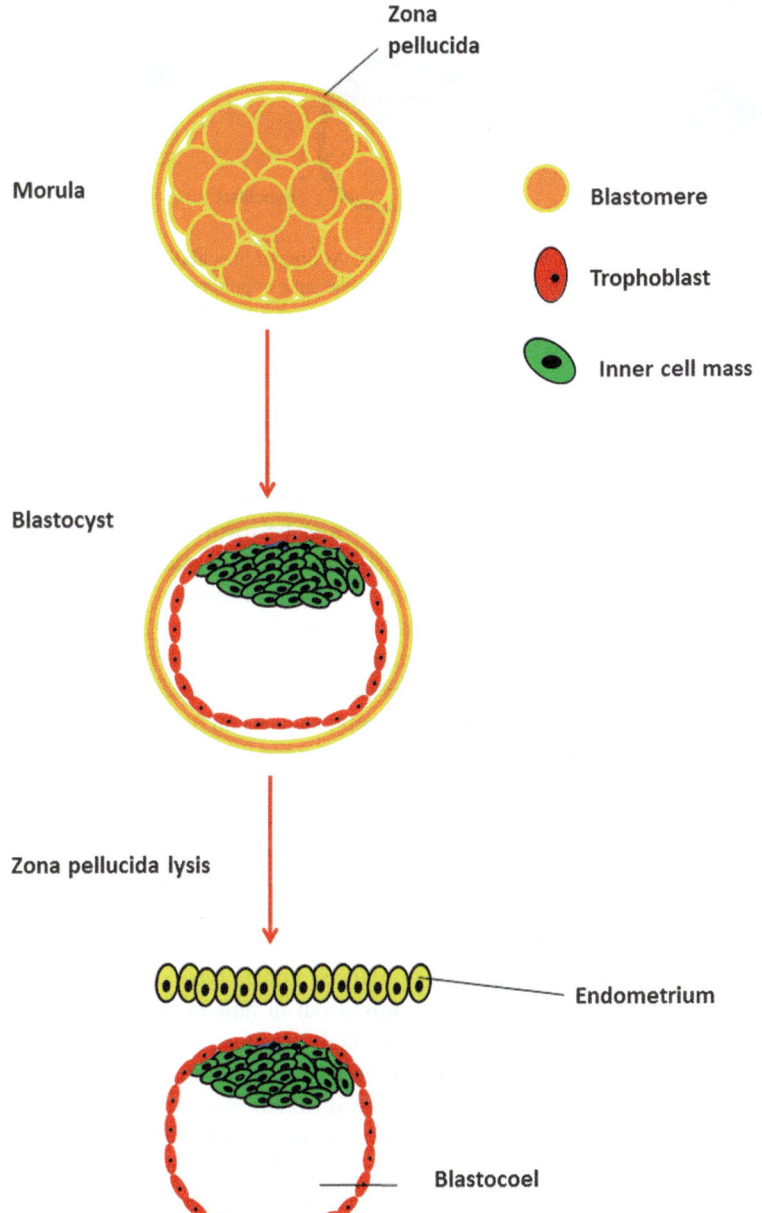

Fig. 1.7 Early embryo development and hatching. Trophoblast cells release a fluid into the central cavity of the blastocyst, forming the blastocoele. The inner cell mass cells move toward one pole and, finally, the zona pellucida is lysed (hatching), enabling the trophoblast cells to directly bind to the endometrium of uterine cavity

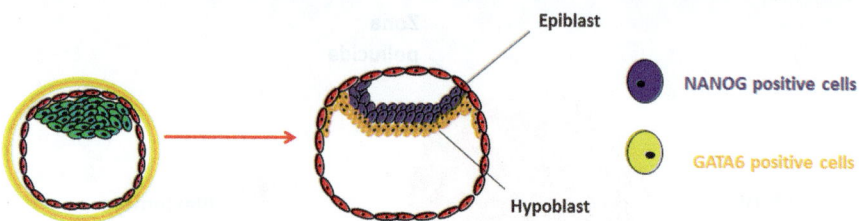

Fig. 1.8 Inner cell mass (ICM) cell differentiation. At time of blastulation, ICM pluripotent cells differentiate into epiblast (NANOG-expressing cells) and hypoblast (GATA6-expressing cells) cells

to a specialized tissue. At present the major goal of developmental biology is to understand how a particular cell differentiates into its final cell type. In this regard, cell proliferation, cell movement, cell specialization, and cell interaction are considered to have key roles and have been shown to be involved in cell commitment, specification, and determination. It has also been demonstrated that each cell constituting an embryo retains a memory of its own cell proliferation history and its positional changes. These complex interactions are regulated by epigenetic modification and differential gene expression that gradually limit cell potency to a phenotype-related expression pattern. These processes have been described in the famous Waddington landscape where a small ball represents a cell of an embryo committing to a certain cell fate by rolling from a noncommitted, pluripotent condition down a hill marked by slopes and valleys. Those (slopes and valleys) address the ball along a progressively more restricted potency pathway, toward a favored position at the bottom of the hill, where the cell is unipotent and is characterized by a tissue-specific differentiated state. This model proposes a "developmental canalization" that allows an organism to develop from the fertilized egg. The entire set of genes expressed by the differentiating organism and their interactions lead to the composition of a "developmental system" that produces a phenotype.

Interestingly, many recent studies carried out in human and murine stem cells have shown that differentiated cells of an adult organism retain a memory of their own cell differentiation history and can be forced in an upstream, countercurrent direction up the differentiation hill, along different states of increased potency. However, it is important to remember that to achieve these results overexpression or activation of additional factors is needed; this leads to the production of induced pluripotent stem cells (iPSCs) that may represent an unlimited source of autologous pluripotent cells, eliminating the immune rejection risk. In contrast, the permanent integration of viral vectors into the host genome to generate iPSCs poses a severe limit to their current therapeutic use. This limitation has stimulated the development of several protocols for a virus-free iPSC derivation, but at present, these approaches are generally more technically demanding and less efficient. Moreover, a recent study carried out in humans has described an alternative method for reprogramming human cells, increasing cell potency through a brief demethylation step achieved with DNA methyltransferase inhibitor 5-aza-cytidine (5-aza-CR).

In vivo and in vitro cell commitment and differentiation are driven by cell-intrinsic properties that control the specification process through an asymmetrical cleavage that leads to an unequal distribution of cytoplasmic determinants (proteins, mRNA, etc.). At the same time, cell specification needs cell-extrinsic signals that derive from cell-to-cell interactions as well as from soluble molecules, known as morphogens, that carry inputs for cell differentiation control.

The major morphogen families directly involved in the formation of specific concentration gradients that drive cells to their correct spatial position and that are fundamental for cell induction toward a specific lineage are as follows.

- *Fibroblast growth factor (FGF) family*: Multifunctional proteins with a wide variety of effects. These proteins are critical during normal development of both vertebrates and invertebrates and have regulatory, morphological, and endocrine effects. The functions of FGF proteins in developmental processes include mesoderm induction, anteroposterior patterning, limb development, and, in mature tissues/systems, angiogenesis and keratinocyte organization. They are also important in neurogenesis, axon growth, and differentiation during development of the central nervous system, and they promote endothelial cell proliferation and the physical organization of endothelial cells into tube-like structures. Thus, FGFs promote angiogenesis, the growth of new blood vessels from the preexisting vasculature; and stimulate repair of injured skin and mucosal tissues by stimulating the proliferation, migration, and differentiation of epithelial cells;
- *Hedgehog (HH) family*: The hedgehog signaling pathway transmits information to embryonic cells required for proper development. It is present at different concentrations in diverse parts of the embryo and is involved in the developmental pattern formation of various organs, such as the eye, brain, gonad, muscle, heart, lung, gut, and trachea. Hedgehog signaling has also been implicated in the development of several human cancers.
- *Wingless (WNT) family*: Secreted lipid-modified signaling glycoproteins that act as ligands to activate the different Wnt pathways via paracrine and autocrine routes. This family has a key role in body axis formation, particularly the formation of the anteroposterior and dorsoventral axes. It is also involved in cell proliferation, migration, and differentiation to prompt formation of organs such as the lungs and ovaries. Furthermore, modulation of Wnt signaling under specific cellular influences can either promote or prevent the early and late stages of apoptotic cellular injury in neurons, endothelial cells, vascular smooth muscle cells, and cardiomyocytes;
- *Transforming growth factor-β (TGF-β) family*: Proteins that control proliferation, cellular differentiation, and other functions in most cells. This family has different important functions related to osteoblast differentiation, neurogenesis, ventral mesoderm specification, angiogenesis, extracellular matrix neogenesis, immunosuppression, apoptosis induction, gonad growth, placenta formation, and left–right axis determination.

1.4   Naïve Versus Established Epiblasts

Embryonic stem cells (ESC) can be captured in vitro in two distinct states of pluri-potency known as "naïve" and "primed" (Fig. 1.9).

Naïve cells can be integrated into a blastocyst and can be converted efficiently to germline chimeras. Furthermore, these cells can be obtained from mouse strains of nonpermissive genetic background and can be modified genetically by homologous recombination.

Recent data showed that it is possible to capture efficiently by including specific cytokines during the establishment of new cell lines from mouse blastocysts or con-vert primed cells to the naïve state by switching culture conditions after the disso-ciation of epiblasts into single cells. In contrast, in nonrodent species naïve ESC have not yet been isolated from preimplantation embryos. Stem cell lines derived from pig embryos share features of primed pluripotency. Furthermore, ESC derived from domestic animals cultured under strictly conditions to force the conversion to a naïve state are highly unstable.

It is widely acknowledged that true ESC can only be derived from mouse and nonhuman primate embryos. Cell lines derived from these species share some major properties such as unlimited replication in vitro (self-renewal), expression of core pluripotency factors such as OCT4, SOX2, and NANOG, capacity to differentiate into any of the different tissues that form the body, and formation of teratomas if injected into immunodeficient mice. However, recently, following the derivation of primate ESC, it became progressively clearer that they present substantial differences when compared to mouse ESC (mESC), even if they share the same definition of ESC.

Differences begin in the derivation of cells. In fact, mouse ESC can be derived only from a small number of 'permissive' strains, whereas primate ESC show no limitations related to a specific genetic background (Fig. 1.10). Mouse ESC require

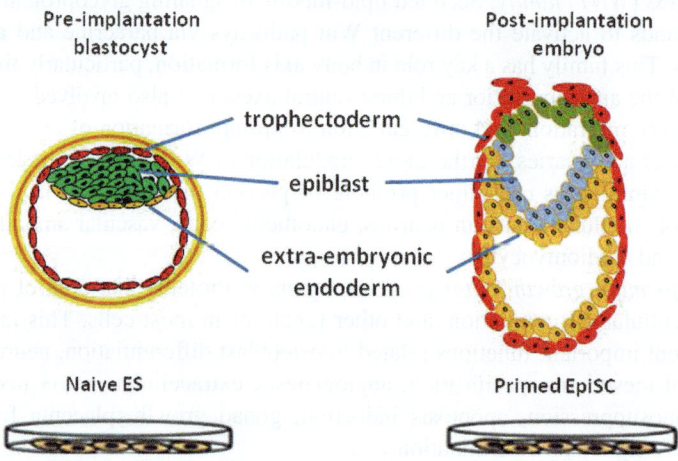

Fig. 1.9 Origin of naïve and primed pluripotent stem cells

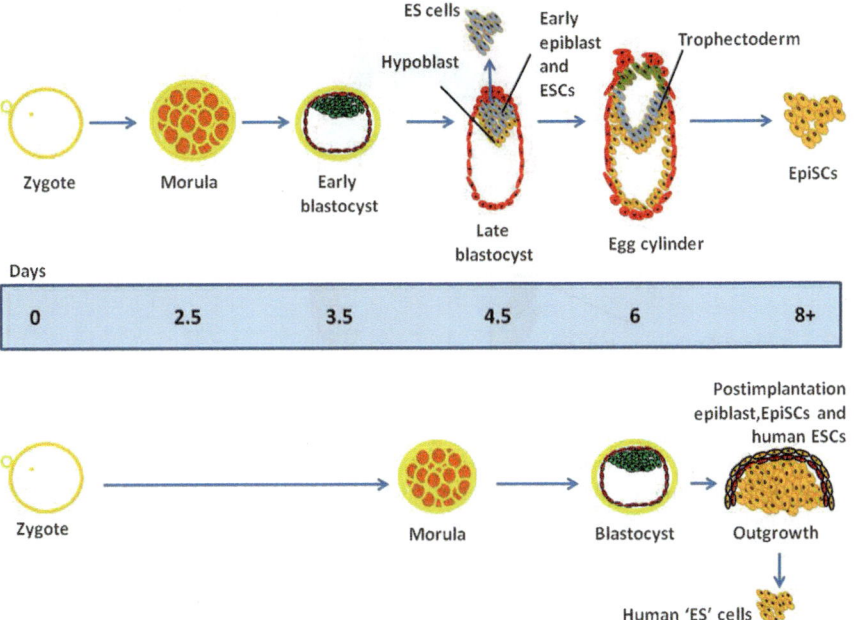

Fig. 1.10 Comparison of pluripotent cell line derivation protocols for mouse and primate embryos. This image highlights the difference in timing of mouse and human development in vivo. It is important to note that the long period of culture that is required for the appearance of human embryonic stem cells would allow explanted cells to progress in vitro to the equivalent of the post-implantation mouse embryo, from which epiblast stem cells (EpiSCs) are derived

Table 1.4 Different origin of naïve and primed stem cells

Naïve	Primed
Mouse preimplantation ICM	Mouse postimplantation epiblast
	Primate preimplantation ICM

leukocyte inhibitory factor (LIF) and bone morphogenetic protein (BMP)-4 in the culture medium to maintain the undifferentiated state of proliferating mouse ESC; primate ESC require activin A or basic fibroblast growth factor (bFGF or FGF2) in the medium because stimulation of the same pathways in primate ESC promotes the differentiation. Moreover, mESC form small, compact, and domed colonies, whereas primate ESC grow in larger and flat colonies. All these differences, and others that are summarized in Table 1.4, have been unexplained for many years, until pluripotent cell lines were derived from the epiblast of postimplantation mouse embryos as opposed to standard preimplantation embryos. These cell lines were named epiblast stem cells (EpiSC). The major property of EpiSC is to share the main characteristics that differentiate human and nonhuman primate ESC from mESC.

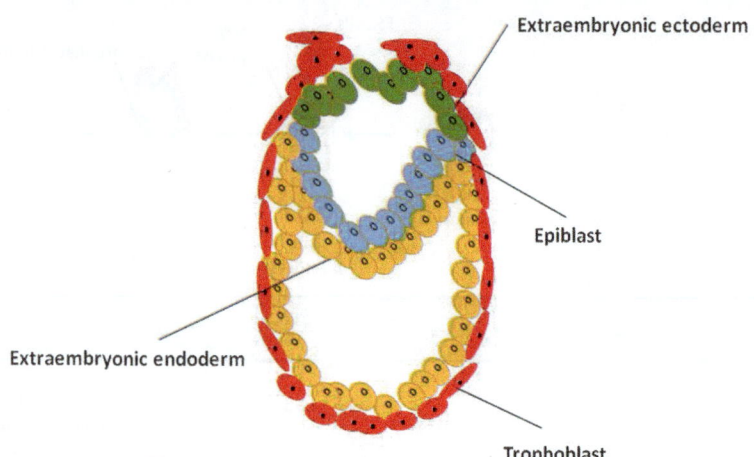

Fig. 1.11 Rodent embryos form the egg cylinder, which requires a thorough reorganization of the epiblast that assumes a cup shape surrounded by hypoblast

ESC and EpiSC

It is now clear that ESC and EpiSC derive from two different stages of embryonic development: one has been defined as a naïve epiblast and can be found in the mouse preimplantation blastocyst, and the other is defined as primed epiblast and is found in nonrodent preimplantation blastocysts and mouse postimplantation embryos. Early embryonic development of rodent species has been extensively studied, and the mouse has been the major model for developmental studies. However, rodent embryos have been shown to possess a distinct and unique aspect that is not common to domestic animals. Following implantation, the mouse embryo forms a polarized and complex structure known as the egg cylinder (Fig. 1.11).

During implantation, the blastocyst hatches from the zona pellucida, attaches the wall of the uterus, and implants. Implantation, in rodents, occurs by a mechanism that is completely different from that operating in the large domestic species. After implantation, the ICM has differentiated into the epiblast and hypoblast (also known as primitive endoderm) and a preamniotic cavity appears in the epiblast (Fig. 1.12).

In the mouse, this is the result of two counteracting signals: a signal inducing apoptosis in epiblast cells and, at the same time, a still unknown survival signal. The latter signal protects the cells in direct contact with the basal lamina from apoptosis and, moreover, ensures that one layer of epiblast cells maintains the pluripotency. These cells now form the embryonic lining of a preamniotic cavity, part of which will later form the amnion, whereas the abembryonic wall of the amnion is formed by so-called amnioblast or extraembryonic ectoderm derived from the upper portion of the epiblast. The mechanism of the formation of the preamniotic cavity in humans and mice is completely different from that operating in the large domestic species.

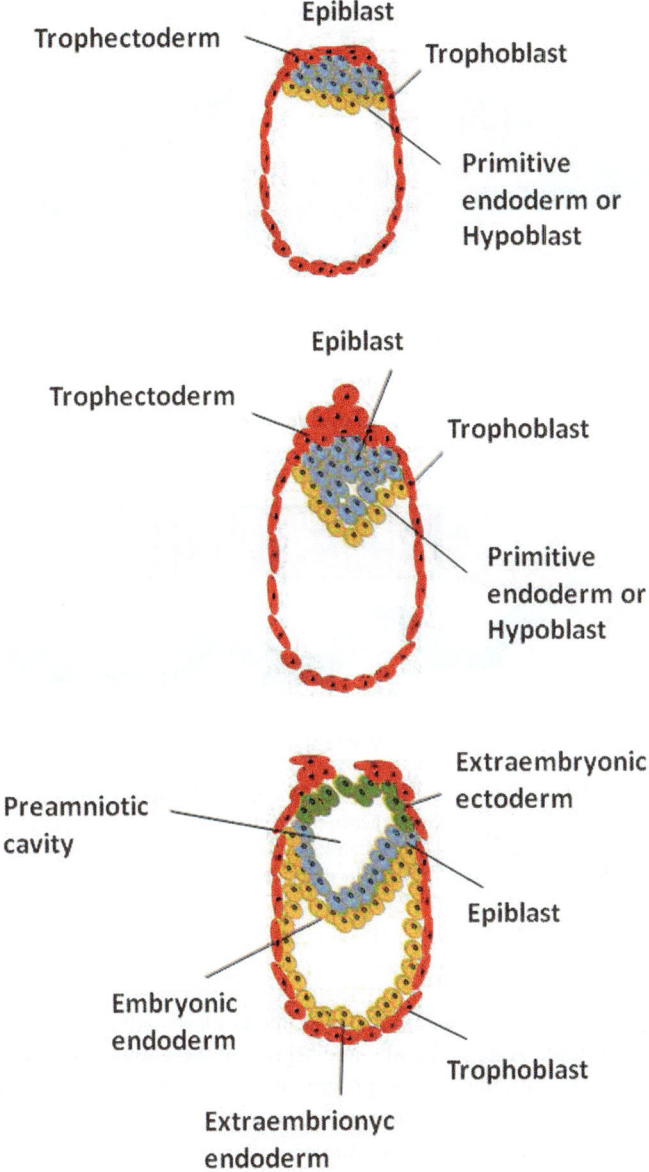

Fig. 1.12 Schematic representation of the development and derivatives of the ICM in mouse embryo. ICM of the murine embryo grows into the blastocyst cavity and forms the egg cylinder

Two aspects of early embryonic development are typical of rodent species: the formation of the egg cylinder and the possibility for early embryos to enter into diapause. Diapause is a phenomenon that can occur in rodents when embryos are

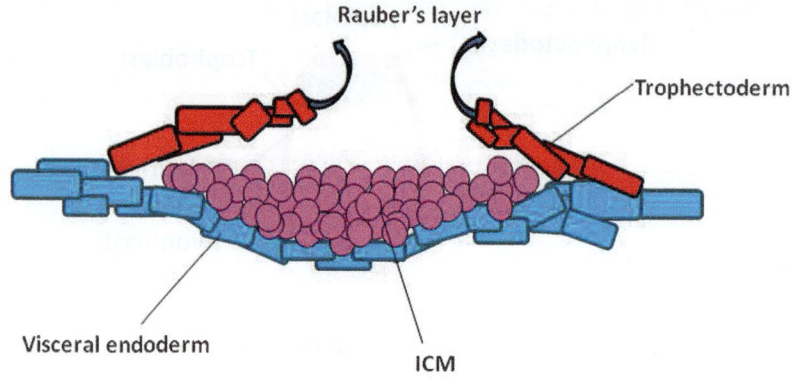

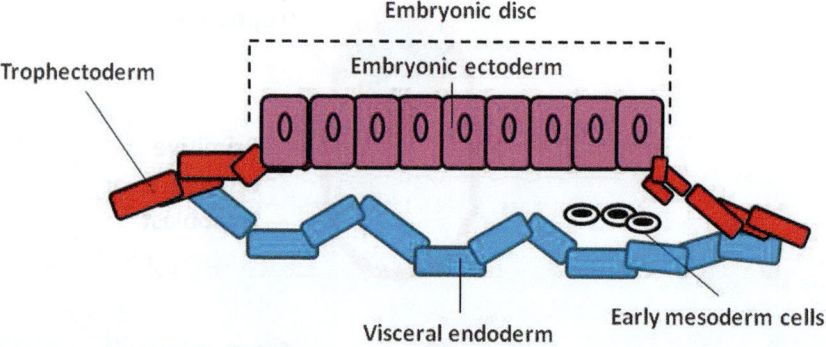

Fig. 1.13 Scheme of the embryonic disc in nonrodent blastocyst. Stage of the loss of polar trophectoderm (Rauber's layer) and of the emergence of the embryonic disc. After blastocyst formation in mammalian embryos, the epiblast delaminates, originating the hypoblast and together they form a flattened multilayered structure known as the embryonic disc

produced in a suckling mother. In diapause, the embryos advance to the blastocyst stage, hatch from the zona pellucida, and segregate the epiblast and hypoblast, but they remain in an unimplanted, nonprogressive state until estrogen is restored. In these circumstances, the epiblast is capable of a low rate of self-renewal, which is supported by LIF secretion by the uterine wall.

In contrast, in domestic animals, around the time of hatching, the ICM differentiates into two cell populations: those facing the blastocyst cavity become flattened and delaminate, forming a inner cell sheet referred to as the hypoblast; and the remaining cells form the multilayered epiblast. In domestic species, the polar trophectoderm covering the epiblast (known as Rauber's layer) gradually disintegrates and is lost, exposing the epiblast to the uterine environment (Fig. 1.13). These differences seem to be among the reasons why it is difficult to derive embryonic stem cells in domestic animal species.

Table 1.5 Different basic properties in naïve and in primed stem cells

	Naïve	Primed
Cell line	Rodent embryonic stem cells (ESC)	Human ESC and rodent epiblast stem cells (EpiSC)
In vitro pluripotency	+	+
Colony morphology	Domed	Flattened
Self-renewal	Rapid	Slow
Teratomas	+	+
Pathway to maintain pluripotency	LIF/BMP-4	FGF-2/activin
Chimera formation	+	−
Single-cell dissociation	+	−
Clonogenicity	+	−

Naïve cells require the presence of leukocyte inhibitory factor (LIF) and bone morphogenetic protein (BMP)-4 in the culture medium to maintain their undifferentiated state. By contrast, primed stem cells need activin A and/or basic fibroblast growth factor (bFGF or FGF2) in the culture medium. Naïve and primed differ also in their morphology: small, compact, and domed colonies are typically formed by naïve cells whereas primed stem cells grow in larger, flat colonies. Naïve colonies are propagated after dissociation to single cells, but the same treatment would rapidly kill primed stem cells whose colonies need to be detached from the feeder layer and fragmented mechanically. One of the most striking properties of naïve stem cells is certainly their capacity to form chimeras when injected into a host blastocyst. The generation of chimeras has been attempted several times with primed stem cells but has never been achieved

ESC Cells in Domestic Animals

Since pluripotent cells were first derived from the ICM of mouse blastocysts, attempts to establish ESC lines have been tried from various mammal species, including pigs, cattle, rats, primates, and humans. Several studies have been conducted with the aim of establishing ESC lines from porcine embryos of various origins. However, bona fide pluripotent stem cells from this species are not yet available, and a much better defined conceptual framework has emerged thanks to the recognition of the difference between naïve and primed epiblast stage and between ESC and EpiSC. Within this framework, it looks as if cell lines belong to the EpiSC type and as such they can sustain a robust self-renewal and form teratomas;their derivation is not restricted by the genotype but is not prone to chimera formation (Table 1.5). Recently, several studies have reported the induction of the pluripotent state in establishment of porcine somatic cells using reprogramming factors. All these cell lines displayed characteristics attributed to primed pluripotent state cells, with flattened morphology and FGF and nodal/activin signaling pathways similar to mouse EpiSC and human ESC. Because it is possible to convert mouse EpiSC into ESC by simply exposing EpiSC to ESC culture medium, it will be interesting to see if EpiSC derived from embryos of domestic animals, for example, the pig, will show the same plasticity and will provide a reliable source of ESC. Recently, studies reported pluripotent stem cells derived from porcine embryos using various exogenous factors.

It will be interesting in the future to see if other methods can be developed to reprogram ungulate EpiSC into naïve ESC working on the epigenome.

Further Reading

Alberio R, Perez AR. Recent advances in stem and germ cell research: implications for the derivation of pig pluripotent cells. Reprod Domest Anim. 2012;47 suppl 4:98–106. doi:10.1111/j.1439-0531.2012.02062.x.

Brevini T, Pennarossa G, Maffei S, Gandolfi F. Pluripotency network in porcine embryos and derived cell lines. Reprod Domest Anim. 2012. doi:10.1111/j.1439-0531.2012.02060.x.

Chen L, Wang D, Wu Z, Ma L, Daley GQ. Molecular basis of the first cell fate determination in mouse embryogenesis. Cell Res. 2010;20(9):982–93.

Coucouvanis E, Martin GR. BMP signaling plays a role in visceral endoderm differentiation and cavitation in the early mouse embryo. Development. 1999;126(3):535–46.

Feldman B, Poueymirou W, Papaioannou VE, DeChiara TM, Goldfarb M. Requirement of FGF-4 for postimplantation mouse development. Science. 1995;267(5195):246–9.

Ferrell Jr JE. Bistability, bifurcations, and Waddington's epigenetic landscape. Curr Biol. 2012; 22(11):R458–66.

Gandolfi F, Pennarossa G, Maffei S, Brevini T. Why is it so difficult to derive pluripotent stem cells in domestic ungulates? Reprod Domest Anim. 2012. doi:10.1111/j.1439-0531.2012.02106.x.

Gilbert SF. Developmental biology. 6th ed. Sunderland: Sinauer Associates; 2000.

Hanna JH, Saha K, Jaenisch R. Pluripotency and cellular reprogramming: facts, hypotheses, unresolved issues. Cell. 2010;143(4):508–25.

Hemberger M, Dean W, Reik W. Epigenetic dynamics of stem cells and cell lineage commitment: digging Waddington's canal. Nat Rev Mol Cell Biol. 2009;10(8):526–37.

Kuijk EW, van Tol LTA, van de Velde H, Wubbolts R, Welling M, Geijsen N, Roelen BAJ. The roles of FGF and MAP kinase signaling in the segregation of the epiblast and hypoblast cell lineages in bovine and human embryos. Development. 2012;139(5):871–82.

Manandhar G, Schatten H, Sutovsky P. Centrosome reduction during gametogenesis and its significance. Biol Reprod. 2005;72(1):2–13.

Nichols J, Smith A. Naive and primed pluripotent states. Cell Stem Cell. 2009;4(6):487–92. doi:10.1016/j.stem.2009.05.015.

Nichols J, Smith A. The origin and identity of embryonic stem cells. Development. 2011;138(1): 3–8. doi:10.1242/dev.050831.

Palermo GD, Colombero LT, Rosenwaks Z. The human sperm centrosome is responsible for normal syngamy and early embryonic development. Rev Reprod. 1997;2(1):19–27.

Pennarossa G, Maffei S, Campagnol M, Tarantini L, Gandolfi F, Brevini TAL. Brief demethylation step allows the conversion of adult human skin fibroblasts into insulin-secreting cells. Proc Natl Acad Sci USA. 2013;110(22):8948–53.

Rodriguez A, Contreras DA, Alberio R. Isolation and culture of pig epiblast stem cells. Methods Mol Biol. 2013;1074:97–110. doi:10.1007/978-1-62703-628-3_8.

Roper S, Hemberger M. Defining pathways that enforce cell lineage specification in early development and stem cells. Cell Cycle. 2009;8(10):1515–25.

Rossant J, Tam PP. Blastocyst lineage formation, early embryonic asymmetries and axis patterning in the mouse. Development. 2009;136(5):701–13.

Sathananthan AH, Kola I, Osborne J, Trounson A, Ng SC, Bongso A, Ratnam SS. Centrioles in the beginning of human development. Proc Natl Acad Sci USA. 1991;88(11):4806–10.

Sathananthan AH, Ratnam SS, Ng SC, Tarín JJ, Gianaroli L, Trounson A. The sperm centriole: its inheritance, replication and perpetuation in early human embryos. Hum Reprod. 1996; 11(2):345–56.

Sathananthan AH, Selvaraj K, Girijashankar ML, Ganesh V, Selvaraj P, Trounson AO. From oogonia to mature oocytes: inactivation of the maternal centrosome in humans. Microsc Res Tech. 2006;69(6):396–407.

Schatten H. The mammalian centrosome and its functional significance. Histochem Cell Biol. 2008;129(6):667–86.

Schatten H, Sun QY. The role of centrosomes in fertilization, cell division and establishment of asymmetry during embryo development. Semin Cell Dev Biol. 2010;21(2):174–84.

Senner CE, Hemberger M. Regulation of early trophoblast differentiation – lessons from the mouse. Placenta. 2010;31(11):944–50.

Sun QY, Schatten H. Centrosome inheritance after fertilization and nuclear transfer in mammals. Adv Exp Med Biol. 2007;591:58–71.

Takahashi K. Cellular reprogramming – lowering gravity on Waddington's epigenetic landscape. J Cell Sci. 2012;125:2553–60.

Ying QL, Wray J, Nichols J, Batlle-Morera L, Doble B, Woodgett J, Cohen P, Smith A. The ground state of embryonic stem cell self-renewal. Nature. 2008;453:519–23.

Zeller R, López-Ríos J, Zuniga A. Vertebrate limb bud development: moving towards integrative analysis of organogenesis. Nat Rev Genet. 2009;10(12):845–58.

Zernicka-Goetz M, Morris SA, Bruce AW. Making a firm decision: multifaceted regulation of cell fate in the early mouse embryo. Nat Rev Genet. 2009;10(7):467–77.

Chapter 2
Pluripotency in Domestic Animal Embryos

Abstract The derivation of stem cells in domestic animal species represents a great improvement in developmental biology and could provide a powerful tool for regenerative medicine and genetic engineering. Despite the enormous efforts and work dedicated to the establishment of stem cell lines from large animals, no conclusive results have been obtained so far, and validated lines have yet to be established.

Here we summarize the data presently available on functional key pluripotency-maintaining pathways and stemness-related marker molecules that can be used to reliably investigate pluripotency in domestic animals.

2.1 Expression, Restriction, and Interactions of Molecules Involved in the Control of Cell Potency and Commitment

Epiblast formation has been extensively described in the mouse. During the first embryonic cleavages, blastomeres are developmentally equivalent: they are all totipotent and all express the transcription factor OCT4.

The first differentiation process consists of polarization, which leads to the generation of two main compartments, the trophectoderm (TE) and the inner cell mass (ICM), both starting from their unique totipotent blastomere precursors. This polarization is manifested by the clear restriction of OCT4 expression to ICM cells. More specifically, it has been demonstrated that OCT4 downregulation in the TE is mainly the result of an active repression exerted by the transcription factor caudal type homeobox 2 (CDX2) (Fig. 2.1).

In the mouse, at the end of polarization, TE cells express CDX2 and ICM cells express OCT4. However, this developmental scenario further and dynamically develops with ICM cells that undergo further differentiation, leading to the formation of the hypoblast, which will lose OCT4 expression, and of the epiblast, that will fully retain expression of the pluripotency-related transcription factor. The epiblast is the tissue that will give rise to the three germ layers in vivo or will lead to the

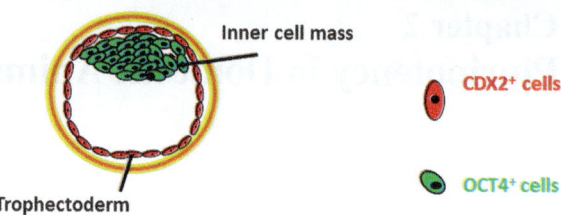

Fig. 2.1 Representative scheme of first polarization. OCT4 expression is restricted to inner cell mass (ICM) cells; CDX2 is transcribed by trophectoderm cells

generation of both primate and rodent ESC when cultured in vitro. Epiblast differentiation and OCT4 restriction to this tissue in the mouse are fully accomplished by day 3.5 of embryo development.

Human embryos go through the same modifications but following slower kinetics. According to Rossant (2011), OCT4 restriction to the epiblast is completed by day 6.

The first observations in domestic animals were carried out by our group in the late 1990s. When we examined the distribution of OCT4 in bovine embryos, we soon realized that the transcription factor was not as tightly restricted to ICM as described in mouse and human embryos but was ubiquitously expressed also in expanded blastocysts. Although these results were quite unexpected, they were the first indication of a different and distinct restriction regulatory mechanism in domestic species. Our early studies were, several years later, confirmed and extended to later-stage embryos. In particular, it was determined that OCT4 restriction to the epiblast is completed only by day 11 in bovine and by day 8–9 in pig embryos. In cattle, hypoblast development is completed by day 10 but the overlying trophoblast, called the Rauber layer, is eliminated a day later, and the epiblast is exposed to the uterine lumen.

This sequence is evidently very different from what is described in the mouse and human. As a consequence, many assumptions that were drawn in domestic animals, based on murine biology, resulted in very misleading concepts and theories in domestic animal gene function.

It has been reported that that the upstream regulatory region of bovine OCT4 is dissimilar from that of mouse and, differing from the mouse, a short sequence is sufficient to counteract CDX2-induced downregulation of OCT4 in the bovine epiblast. This observation is supported by the experiments recently carried out, wherein fusing the bovine OCT4 upstream region with the mouse gene resulted in the prevention of OCT4 restriction to the ICM and allowed expression of the transcription factor in the TE as well.

Epiblast/hypoblast differentiation from their common ICM precursor reflects the differences just mentioned and related to the various species. In the mouse, this process is regulated by fibroblast growth factor (FGF), which activates a mitogen-activated protein kinase and results in the upregulation of GATA6 expression while repressing the transcription of Nanog. In humans, the mechanisms are less well understood. FGF does not seem to have an effect; as a consequence, it is unclear how epiblast formation is regulated. In cattle, a further regulatory mode is likely to take place, where FGF inhibits NANOG but has no control of GATA6 expression.

It remains an open question whether these significant species variations exert a limiting effect on the possibility of deriving domestic animal pluripotent cell lines.

2.2 Species-Specific and Stage-Specific Pluripotency Markers

Can we do without OCT4?

Classic molecules accepted as pluripotency markers in human and mouse ESC are still not the gold standard for large animal species. OCT4, SSEA1, SSEA4, and alkaline phosphatase are indeed expressed by ungulate ICM and embryo-derived cell lines. However, the same genes are also expressed in the trophectoderm, which greatly hampers an accurate and reliable evaluation of putative cell lines

For example, the OCT4 expression mode in the porcine model is debated. Many reports demonstrate that both ICM and trophectoderm express this molecule. On the other hand, other laboratories show an inner cell mass-confined restriction of OCT4 in embryos at the expanding hatched blastocyst stage.

In our studies we find that OCT4 mRNA is present at the time of porcine ICM plating and during the following passages. However, by passage seven to ten, its expression is completely downregulated or, when expression persists, immune positivity extends to the cytoplasmic compartment and is not only restricted to the nucleus (Fig. 2.2).

One interesting aspect to keep in mind is that this downregulation of OCT4 does not appear to affect cells, because when we cultured them for several further months, no signs of changes in their morphology and no expression of specific differentiation markers were present.

Altogether, these findings imply that, even though OCT4 is likely to be a marker of stemness in the pig also, it does not appear to be either the only or the key transcription factor playing a pivotal part in the maintenance of pluripotency in this species. One hypothesis that may explain these observations is that OCT4 might exert an indispensable effect in plating and early culture of pig epiblast, but may then be replaced by other pluripotency factors such as NANOG, which is another well-characterized marker for human and mouse stemness.

NANOG is consistently expressed in porcine cell lines and is reported to be strongly downregulated in caprine trophectoderm while being strongly expressed in the ICM (Fig. 2.3). It also appear to be a specific marker of pluripotency for ruminants because both its mRNA and protein are found in the ICM and strongly downregulated in the trophectoderm

Further understanding is needed, but we propose that NANOG may be able to maintain pluripotent cells in an undifferentiated state, and also in the absence of the simultaneous expression of OCT4.

On the other hand, we may hypothesize that NANOG expression, in the absence of OCT4, indicates a "stand-by" mode, wherein a cell is prevented from committing

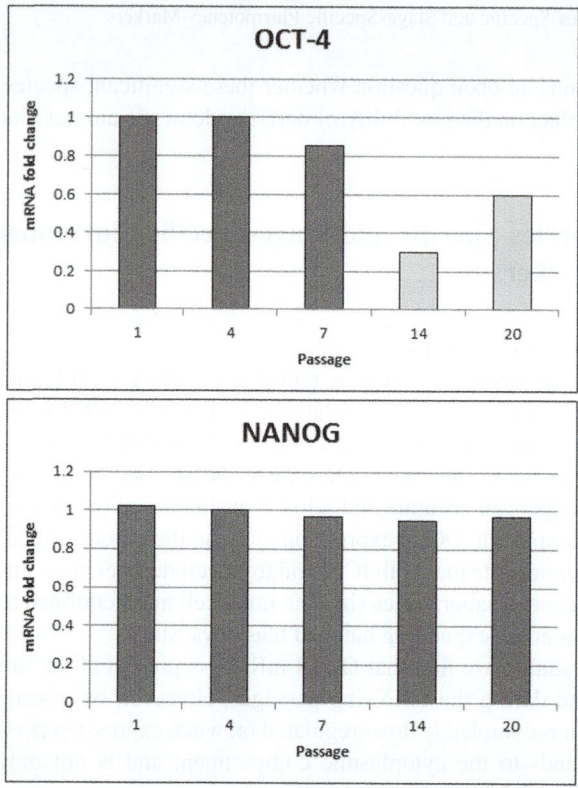

Fig. 2.2 OCT4 and NANOG mRNA expression in porcine cultured ICMs. OCT4 mRNA is present at the time of ICM plating; however, by passage seven to ten, its expression is completely downregulated. By contrast, NANOG is consistently expressed in porcine cell lines

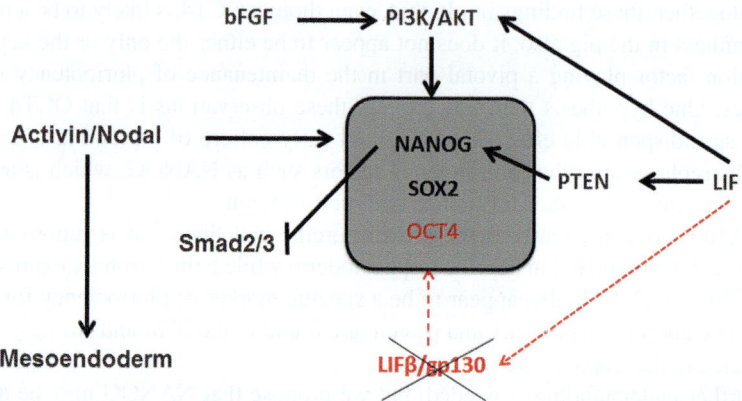

Fig. 2.3 Scheme of leukemia inhibitory factor (LIF), fibroblast growth factor (FGF), and activin pathways in pig. These molecules act in the pig through a cascade that involves phosphoinositide-3-kinase (PI3K), serine/threonine protein kinase (AKT) (a key effector in the PI3K pathway), and phosphatase and tensin homologue deleted on chromosome 10 (PTEN). The activin/nodal signaling pathway controls the maintenance of pluripotency, controlling the expression of NANOG, which in turn limits the transcriptional activity of the Smad2/3 cascade, and blocking progression along the activin/nodal-dependent mesoendoderm differentiation

Fig. 2.4 The presence of LIF
in the culture medium
inhibits the differentiation
process and prevents
embryoid body formation

WHITOUT LIF

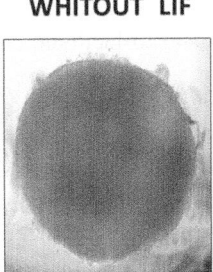

WITH LIF

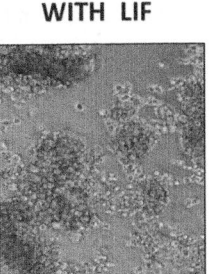

to differentiation but at the same time is not fully pluripotent; the simultaneous expression of both factors (and possibly many others) is required to maintain cells in a truly pluripotent state. Certainly the dispensability of OCT4 in pigs has been recently shown with experiments that allowed for the derivation of pig induced pluripotent stem cells (iPS) with a protocol that only used three reprogramming factors and did not take advantage of the forced expression of OCT4.

Experience from our laboratory and others showed that pluripotency regulation/ repression via the LIF/JAK/STAT3 pathway is very debatable in porcine cell lines (Fig. 2.3) and suggests that these cells are likely to belong to the epiblast stem cell (EpiSC) type. This finding indicates that the pig, and possibly other ungulates, responds to culture conditions in a way that is more similar to primate embryos than to murine embryos. On the other hand, although we have found that pig cell lines do not express leukemia inhibitory factor (LIF) receptor, indicating that the addition of this factor to the culture medium is not essential for the maintenance of pluripotency, the cytokine appears to inhibit the differentiation process because its presence in the standard medium used for embryoid body (EB) formation results in preventing cell commitment to germ layer specification and inhibited cell aggregation (Fig. 2.4).

One possibility that cannot be ruled out is that LIF might exert its role in pluripotency maintenance through a nonclassical LRβ-gp130 and STAT3 activation pathway.

Also, LIF might exert its effect through alternative signaling pathways that have been shown to participate in maintaining pluripotency.

Pluripotent porcine cells have been isolated and maintained in culture in a medium containing either LIF and FGF2 or a very complex mixture of LIF, FGF2, activin, and epidermal growth factor (EGF). Therefore, the fundamental role of each of these molecules remains obvious.

When we examined the expression pattern of some of the genes involved, we could observe that the receptor for FGF2 is constantly expressed in our porcine cells, suggesting that pig stem cells may have culture needs more similar to those of primate cells than to rodent cells.

A closer look gave us further interesting information. In particular, we could demonstrate the expression of PI3K, AKT (a key effector in the PI3K pathway), and PTEN (a negative regulator of the same pathway). The PI3K/AKT signaling cascade is known to be responsive to LIF and has been previously shown to trigger the expression of NANOG and to facilitate efficient proliferation and survival of murine ESC.

This observation is not inconsistent with our observation on the role of FGF2 because this molecule can also bind and activate the PI3K/AKT cascade. Therefore, the good results obtained by different laboratories with the combined use of FGF2 and LIF may find an explanation when we consider that the two molecules can act synergistically to promote self-renewal through common pathways.

Very interesting results have been recently obtained using day 10–12.5 elongated pig blastocysts as a source (late, or primed epiblast) and culturing in a medium enriched with FGF2 and activin A. These stem cell lines, defined as pEpiSC, can be cultured for extensive periods and can be induced to differentiate to three somatic germ layers, germ cell progenitors, and trophoblast. This study indicated the involvement of the FGF/activin/nodal signaling pathway for maintenance of pluripotency in mammals.

All these important observations related to the stem cell can be exploited for developing strategies addressed to the derivation of pluripotent ESC from ungulate embryos rather than to LIF, therefore, with cell lines showing a robust self-renewal and the ability to differentiate into precursor cells derived from all three germ layers as well as into trophectoderm and germ cell precursors.

Further Reading

Alberio R, Croxall N, Allegrucci A. Pig epiblast stem cells depend on activin/nodal signaling for pluripotency and self-renewal. Stem Cells Dev. 2010;19(10):1627–36.

Berg DK, Smith CS, Pearton DJ, Wells DN, Broadhurst R, Donnison M, Pfeffer PL. Trophectoderm lineage determination in cattle. Dev Cell. 2011;20(2):244–55.

Brevini AL, Tosetti V, Crestan M, Antonini S, Gandolfi F. Derivation and characterization of pluripotent cell lines from pig embryos of different origins. Theriogenology. 2007;67(1):54–63.

Brevini TA, Pennarossa G, Attanasio L, Vanelli A, Gasparrini B, Gandolfi F. Culture conditions and signalling networks promoting the establishment of cell lines from parthenogenetic and biparental pig embryos. Stem Cell Rev. 2010;6(3):484–95. doi:10.1007/s12015-010-9153-2.

Feldman B, Poueymirou W, Papaioannou VE, DeChiara TM, Goldfarb M. Requirement of FGF-4 for postimplantation mouse development. Science. 1995;267(5195):246–9.

Gandolfi F, Pennarossa G, Maffei S, Brevini TAL. Why is it so difficult to derive pluripotent stem cells in domestic ungulates? Reprod Domest Anim. 2012;47(5):11–7.

Hall VJ, Christensen J, Gao Y, Schmidt MH, Hyttel P. Porcine pluripotency cell signaling develops from the inner cell mass to the epiblast during early development. Dev Dyn. 2009;238(8):2014–24.

Kuijk EW, van Tol LTA, van de Velde H, Wubbolts R, Welling M, Geijsen N, Roelen BAJ. The roles of FGF and MAP kinase signaling in the segregation of the epiblast and hypoblast cell lineages in bovine and human embryos. Development. 2012;139(5):871–82.

Rossant J. Developmental biology: a mouse is not a cow. Nature. 2011;471(7339):457–8.

Talbot NC, Blomberg LA. The pursuit of ES cell lines of domesticated ungulates. Stem Cell Rev. 2008;4(3):235–54.

van Eijk MJT, van Rooijen MA, Modina S, Scesi L, Folkers G, van Tol HTA, Bevers MM, Fisher SR, Lewin HA, Rakacolli D, Galli C, de Vaureix C, Trounson AO, Mummery CL, Gandolfi F.

Molecular cloning, genetic mapping, and developmental expression of bovine POU5F1. Biol Reprod. 1999;60(5):1093–103.

Vassiliev I, Vassilieva S, Beebe LF, Mcllfatrick SM, Harrison SJ, Nottle MB. Development of culture conditions for the isolation of pluripotent porcine embryonal outgrowths from in vitro produced and in vivo derived embryos. J Reprod Dev. 2010;56(5):546–51.

Vejlsted M, Du Y, Vajta G, Maddox-Hyttel P. Post-hatching development of the porcine and bovine embryo: defining criteria for expected development in vivo and in vitro. Theriogenology. 2006;65(1):153–65.

Chapter 3
Use of Large Animal Models for Regenerative Medicine

Abstract Regenerative medicine requires preclinical trials of new therapies before starting human studies. In this context, animal models play a fundamental role for investigating biological and functional activities of new cells and tissues. The rodent species is extensively used in studies related to stem cell biology, providing important information, but at the same time its use has important limitations for a variety of disease categories because of the different body size and physiology relative to humans. Large animal species, such as dogs, pigs, sheep, cattle, horses, and nonhuman primates, are better predictors of human responses than rodents, but in each case it will be necessary to choose the best model for a specific application.

The knowledge of embryonic and induced pluripotent stem cells, as well as of adult stem cells, requires significant effort for establishing and characterizing cell lines, comparing behavior to human analogues, and testing potential applications.

Herein we describe the current status and advantages of the use of large animal models in stem cell-based regenerative medicine.

3.1 Why Do We Need Stem Cell Lines from Large Animal Species?

Experimental models provide a fundamental tool to investigate biological and functional cell and tissue activities as well as to study diseases. Animals provide a 'whole organism' to test and perform preclinical trials of new therapies before starting human studies. The use of animal systems also facilitates experimental design by providing controls with matching genetic background, age, gender, and exposure history.

These models are commonly divided in different classes: invertebrates (i.e., flies, worms), small animals (i.e., fish, rodents), and large animals (i.e., swine, primates, sheep).

T.A.L. Brevini and F. Gandolfi, *Pluripotency in Domestic Animal Cells*,
SpringerBriefs in Stem Cells, DOI 10.1007/978-1-4899-8053-3_3, © The Author(s) 2013

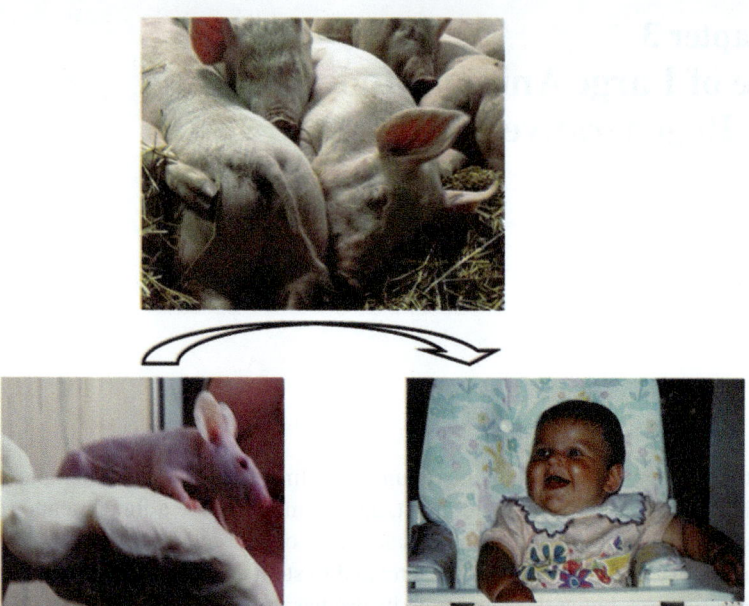

Fig. 3.1 The murine model does not reliably reproduce human diseases because there are many differences between these two species. Thus, the use of large animals is necessary to provide an enhanced tool to predict clinical efficiency

As in many other areas of research, the murine species is extensively used in studies related to stem cell biology. However, this model does not reliably reproduce human diseases, and there are many differences between these two species. This problem represents an important limitation to using mouse embryonic stem cells (ESC) as models for regenerative medicine. During past years, significant progress has been made in the creation and use of humanized mice, which contain human genes or modified tissues that allow the investigation of human mechanisms, maintaining an in vivo context within the mouse. Nevertheless, the existing differences between donor and recipient animals continue to affect the survival of transplanted stem cells because of species differences in the trophic properties of tissues.

In this context, the use of large animals as models becomes necessary because this approach provides an enhanced ability to predict clinical efficiency relative to mice (Fig. 3.1). Moreover, the use of large animals as a model presents different important advantages because these organisms have developed as outbred populations and their immune and physiology systems are closer to those of the human than the mouse.

Obviously, a critical step in this respect is represented by the selection of the animal that is most appropriate for each specific study. Usually the pig is considered as one of the best models to study human diseases because of its well-known similarities in terms of anatomy, physiology, metabolism, and organ development with the human. Furthermore, the creation of humanized pigs has been recently reported, as well as the improvement of preclinical disease models created by targeted genetic engineering; these would be very suitable models for testing stem cell therapies.

3.2   What Are the Applications and Potentials?

Among the different large animal species, mini-pigs and full-size breeds have been widely used for studies such as infectious diseases, cardiovascular disease, atherosclerosis, ophthalmology, digestive processes, and cancers (Table 3.1). The importance of this species as a biomedical model has been enhanced during past years by targeting specific genomic sites for modification. Swine disease models created by targeted genetic engineering include those for cystic fibrosis, Alzheimer's and Huntington's disease, retinitis pigmentosa, hyperlipoproteinemia, and muscular dystrophy.

In our laboratory, we carried out studies for cardiac stem cell therapy and selected porcine species as a model. Similar experiments carried out until now were predominantly performed on mice, but significant differences exist between cardiac characteristics when mice are compared with the human (size, heart rate, coronary architecture, and general anatomy). For this reason, in the particular field of cardiovascular pathologies, large animals such as dogs, sheep, and pigs are proposed as alternative models.

Important limitations of working with sheep and ruminants derive from their gastrointestinal and thoracic anatomy, which are very different from those of monogastric species. In contrast, the dog represents a popular model for myocardial ischemia and infarction studies. The current reperfusion treatment guidelines for acute coronary syndromes used in the human were developed in the dog some years ago. However, during past years, some divergences have been detected and highlighted

Table 3.1   Similarities and differences among human, pig, and mouse

	Body mass (g)	Longevity (years)	Gestation (days)	Heart weight (g)	Lung weight (g)	Liver weight (g)
Human	50,000–90,000	70	280	320	1,200	1,500
Pig	60,000–150,000	25	114	370	1,000	1,500
Mouse	40	1–3	21	0.05	0.15	1.50

The pig is considered the best model to study human diseases because it has well-known similarities with the human in terms of anatomy (body mass; heart, lung, and liver weights) as well as longevity and gestation, whereas many differences are present for the same characteristics between human and mouse

Table 3.2 Comparison of heart anatomy and physiology between human and the most common heart disease animal models

	Mouse	Rabbit	Dog	Sheep	Human	Pig
Heart weight (g)	0.14–0.15	9–11	160–420	240–360	360–480	400–500
Heart rate/min	500–600	120–300	60–120	70–80	60–90	65–75
Systolic pressure (mmHg)	80–160	70–170	120–150	80–120	60–120	70–130

when the dog was compared to the human (heart size, body weight, collateral coronary circulation approximately four times more extended than that of humans). Therefore, the new trend involves the use of the pig as a model.

3.3   A Lesson from the Heart

Most of the experiments carried out until a few years ago in cardiac stem cells and regenerative medicine were predominantly performed on mice and human. This selectivity restricted quite significantly application of the results obtained in preclinical studies that could not be performed using the human as a model and, at the same time, were limited by the evident differences between mouse and human (Table 3.2). In particular, human and mouse hearts diverge in coronary architecture, the variations of which are much greater in humans compared to mice. Indeed, the limited murine variations in heart vessel architecture consequently result in the size and location of the ischemic area being fairly constant in the mouse while being more varied in the human.

Differences can also be appreciated at the cellular level, as indicated by the higher capillary density and larger cross-sectional area of myocytes in humans in comparison to the mouse. Consequently, extrapolation of murine systems, particularly after induction of cardiovascular stress, must be meticulously monitored when applied clinically because of the obvious differences between the two systems.

The available literature demonstrates many similarities between pig and human heart anatomy and shows that the pig represents an optimal model in therapy and cardiovascular research as an intermediate animal, offering the ideal link between the classical rodent models and the human and thus representing an experimental tool that may more readily be used and translated in preclinical approaches. For instance, coronary anatomy and the poor subendocardial to epicardial collateral network of the swine heart are very similar to those of the human.

Studies carried out in the porcine model have confirmed this species to be an optimal model for regenerative medicine after myocardial infarction and have allowed us to demonstrate the presence of resident cardiac progenitors that, although more abundant in early postnatal life, persist and assure local remodeling in the adult organism as well. These cells were isolated from three different heart regions,

Aorta	**Ventricle**	**Atrium**

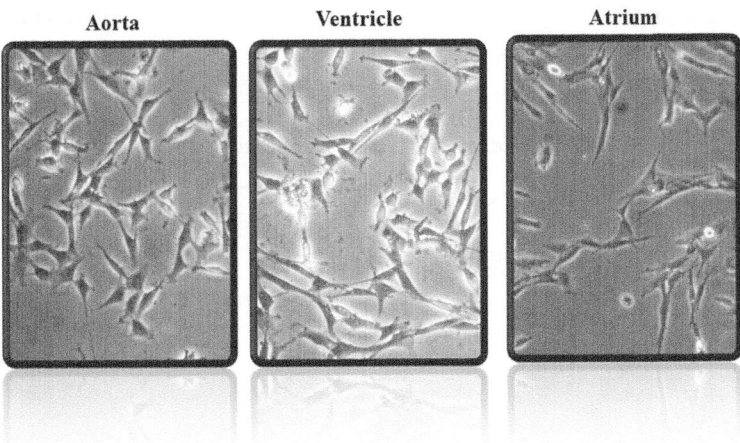

Fig. 3.2 Cardiac progenitor cells isolated from explants of aorta, ventricle, and atrium of a healthy adult pig heart

the aorta, ventricle, and atrium, respectively, indicating that this subpopulation of committed, but still proliferating, cardiac progenitor cells is not confined to a specific region within the organ but is rather distributed to several areas (Fig. 3.2). Interestingly enough, these cells seem to be more abundant in the outer layers of the organ, indicating the epicardium as a possible site of origin for resident stem or progenitor cell populations in the pig.

One limit to the use of these cells in translational experiments is the need for immunosuppression in the pig, which is problematic and economically disadvantageous. However, one promising strategy, the use of syngeneic animals obtained through nuclear transfer, may allow us to overcome the problems connected with immune response reactions (see Fig. 3.3).

Other important similarities between human and pig are represented by blood glucose levels and insulin isoform/effect, which have led to using the pig as a source for clinical pancreatic islet transplantation. Current therapy for diabetes is based on an intensive administration of exogenous insulin for the control of blood glucose levels; nevertheless, patients may often develop complications such as angiopathy, nephropathy, retinopathy, or cardiopathy. Pancreatic islet transplantation for damaged tissue replacement is one of the most attractive and promising strategies for patients. In line with this, recent studies have shown the possibility of using human islets for allogeneic transplantation, but two to three donors are required to cover the needs of a single recipient. In contrast, whole pancreas transplantation requires only one donor, but current pancreas preservation techniques are able to recover less than half of the approximately 1 million islets present in a single pancreas. Xenogeneic islet/tissue transplantation may be a useful tool to overcome the problem of the scarcity of human islets (Fig. 3.4).

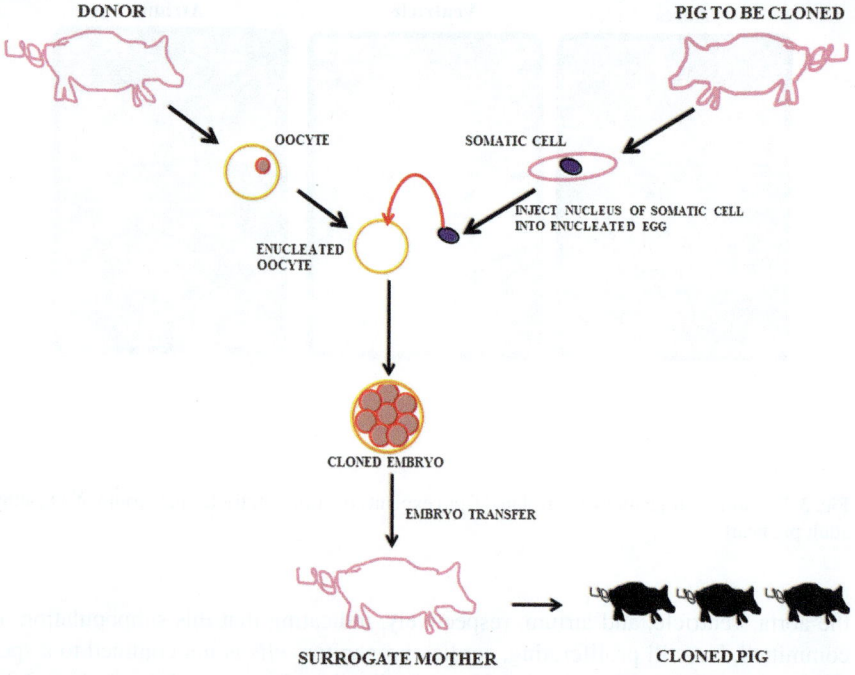

DONOR   **PIG TO BE CLONED**

OOCYTE   SOMATIC CELL

INJECT NUCLEUS OF SOMATIC CELL
INTO ENUCLEATED EGG

ENUCLEATED
OOCYTE

CLONED EMBRYO

EMBRYO TRANSFER

SURROGATE MOTHER   **CLONED PIG**

Fig. 3.3 Diagram of nuclear transfer procedure. This technique allows obtaining syngeneic animals that may be used to overcome the problems connected with immune response reactions

Fig. 3.4 Representative
picture of isolated pancreatic
islets

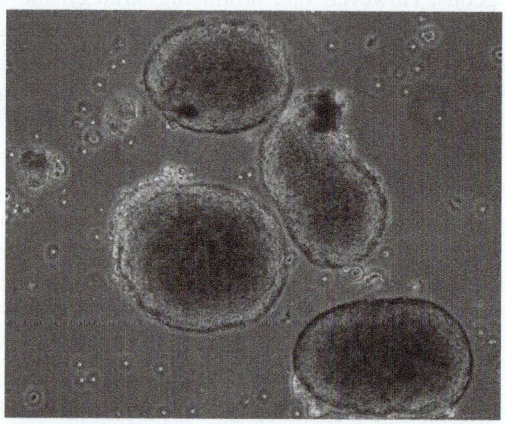

3.4   Therapies with Mesenchymal Stem Cells in the Horse…   Are We There?

A Special Perspective

There are many cell sources for regenerative medicine but it is not clear which option will prove to be therapeutically optimal. At the moment, the best hope for effective different cell lines are mesenchymal stem cells (MSCs), previously known as stromal stem cells.

The term mesenchymal stem cells was coined in 1991 by Arnold Caplan and became widely adopted.

However, not more than a year ago Caplan proposed using the alternative term medicinal signaling cells instead of mesenchymal stem cells. He believes that MSCs are not stem cells, but progenitors. He thinks that cell definition should relate to its functions. MSCs become activated in an inflammatory environment and express their "medicinal" functions, which are, namely, immunomodulatory and trophic. So, in a way the biological function of MSCs has nothing to do with stemness. He supports this idea, sharing that scientists should focus on what these cells can do medicinally instead of how they can differentiate. Furthermore, as the experimental proof whether MSCs or even their subsets fulfill the stem cell definition is still lacking on a single-cell level, the term stem cell seems to be inappropriate. Despite this, if the MSCs qualify as stem cells remains a legitimate question; the term mesenchymal stem cells has gained such global usage that professionals have not yet adopted the new name "medicinal signaling cells."

MSCs or MSC-like cells have been isolated from adult bone marrow, adipose tissue, and amniotic fluid as well as from fetal blood, liver, bone marrow, and lung (Fig. 3.5).

The first source reported to contain multipotent progenitor cells was the stromal compartment of bone marrow. For this reason, bone marrow is currently the best investigated origin of MSCs in domestic animals. MSCs derived from bone marrow (BM-MSCs) can be directed toward the osteogenic, chondrogenic, and adipogenic lineage (Fig. 3.6).

These cells have a different potency of differentiation compared to MSCs derived from other sources. For example, chondrogenic differentiation capability is higher in MSCs derived from bone marrow when compared to MSCs derived from adipose tissue, but was inferior when compared to umbilical cord blood-derived MSCs. Although BM-MSCs represent the cells commonly investigated for application in human and veterinary regenerative medicine, it is important to remember that BM-MSCs have a more limited potential than embryonic stem cells in terms of in vitro proliferation ability. They also show minor plasticity and growth with increasing donor age and in vitro passage number. Adipose tissue is another source of MSC

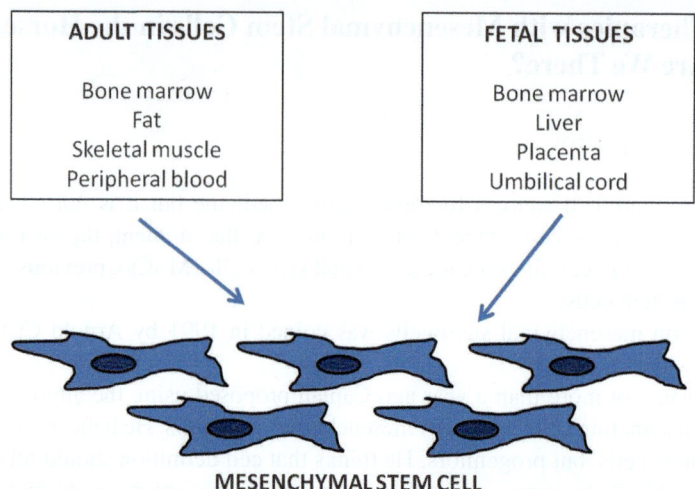

Fig. 3.5 Mesenchymal stem cells (MSCs) can be isolated from adult tissues (bone marrow, fat, skeletal muscle, peripheral blood) and fetal tissues (bone marrow, liver, placenta, umbilical cord)

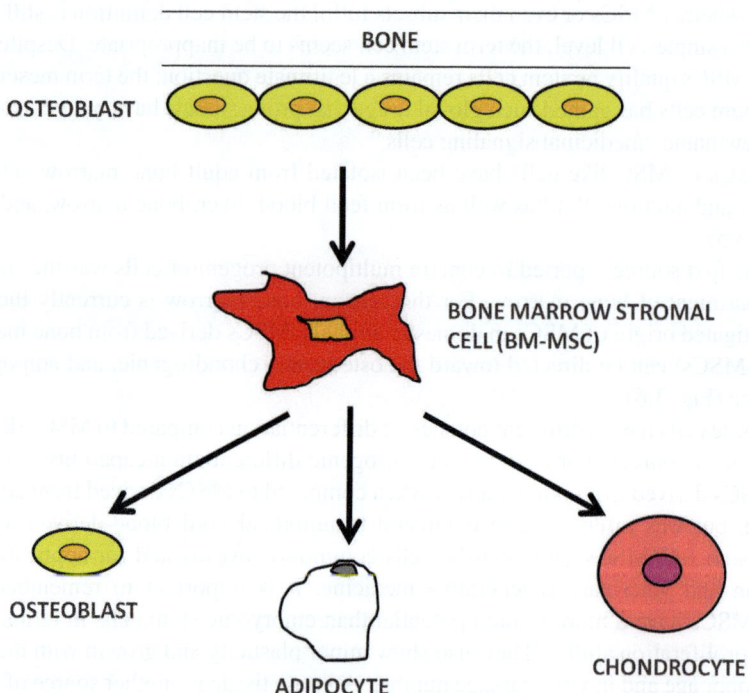

Fig. 3.6 MSCs derived from bone marrow (BM) and multipotent differentiation capacity. BM-MSCs can generate multiple mesoderm-type cell lineages, such as osteoblasts, adipocytes, and chondrocytes

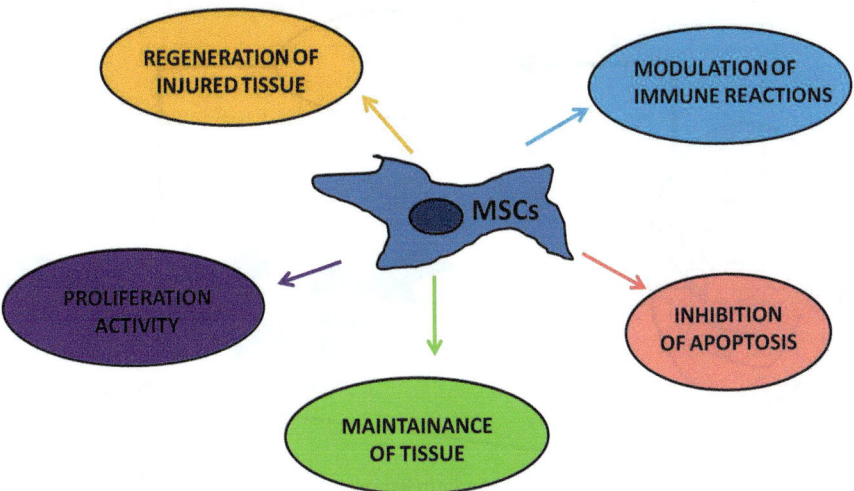

Fig. 3.7 MSCs secrete soluble factors that have beneficial effects on regeneration of injured tissues, inhibit apoptosis, stimulate proliferation, and modulate immune reactions

frequently used in humans and animals. The encouraging results obtained with the recovery of adipose-derived MSC in domestic animals suggests that fat is a promising MSC source for clinical application. In fact, these cells display a high proliferation rate and less senescence compared to MSCs from other sources. Furthermore, MSCs derived from adipose tissue also showed high ability to differentiate into mesodermal lineage.

MSCs have recently received much attention for their therapeutic potential in regenerative medicine.

Several studies have demonstrated the capability of MSCs to secrete soluble factors that have beneficial effects on the regeneration of injured tissues. They were also found to inhibit apoptosis, limit pathological fibrotic remodeling, stimulate proliferation and differentiation of endogenous stem-like progenitors, decrease inflammatory oxidative stress, and modulate immune reactions (Fig. 3.7).

In recent years, a number of studies have reported the results of stem cell therapy in the treatment of domestic animals, focusing on the treatment of arthritis, atopic dermatitis, and tendon injury.

The use of adipose-derived MSCs for the treatment of osteoarthritis was studied in the dog. A mixed population of cells was transplanted with intraarticular injection and, following treatment, the dogs showed a significant improvement in functional ability.

One major focus of regenerative medicine in large animals is the regeneration of tendons and ligaments because several lines of evidence suggest that MSCs are present also in these structures. Implantation of MSCs from different sources in far greater numbers than are present normally within tendon tissue would have potential for regenerating or repairing the tendon. MSCs have been implanted into surgical defects in tendons in multiple in vivo experiments in laboratory animals with mostly positive outcomes. It is still unclear whether the major contribution of the

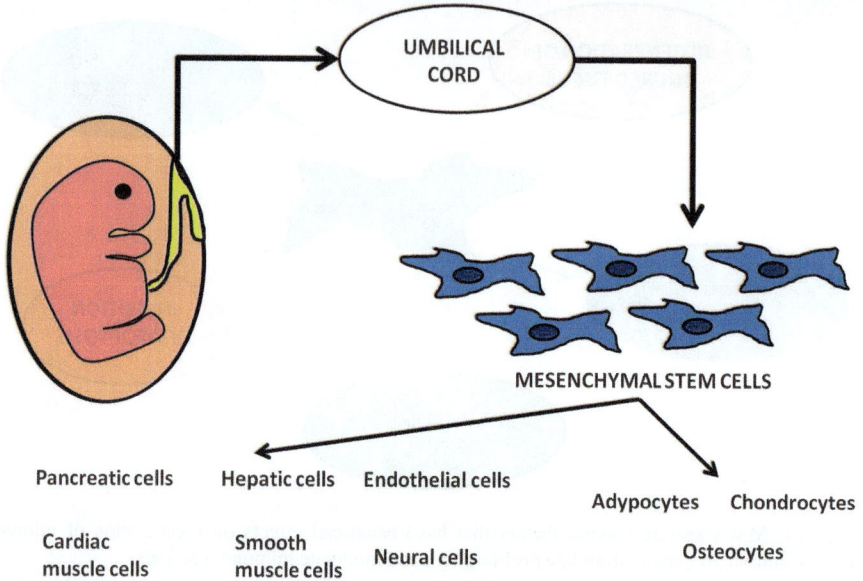

Fig. 3.8 MSCs derived from the umbilical cord have the potential to generate mesodermal lineages (adipocytes, chondrocytes, osteoblasts) and they can also trans-differentiate into some cell lineages from other germ layers (cardiac muscle cells and smooth muscle cells; endothelial cells from mesodermal layer; neural cells from ectodermal layer; hepatic cells and pancreatic cells from endodermal layer)

MSCs to the healing process is the differentiation into tenocytes or rather the supply of growth factors and thus the stimulation of residing cells within the tendon, or whether a combination of the two mechanisms occurs. Several studies on horse tendinopathies revealed that injected cells remain located within injected lesions and that both autologous and allogeneic MSCs may be used without stimulating an undesirable cell-mediated immune response from the host.

The available literature demonstrates the evidence of the benefit and safety of adult MSC application for tissue engineering, an emerging field that offers outstanding opportunities for regenerative medicine. However, the collection of bone marrow and adipose tissue in several animal species, such as the horse, requires an invasive procedure.

To overcome the invasive collection of bone marrow and adipose tissue, progenitor cells derived from extra-fetal sources, such as the umbilical cord and amnion, could represent alternative candidates (Fig. 3.8). In addition to the noninvasive nature of the isolation procedure, the use of adnexal tissue suggests that cells may have immunomodulatory characteristics that could decrease the risk of the recipient rejecting transplanted stem cells immunologically.

The first amnion-mesenchymal cells (AMCs) were derived and characterized in the horse. In these studies, when AMCs were compared to equine BM-MSCs, AMCs and BM-MSCs both exhibited adult stromal cell-specific gene and protein expression but showed differences in density of collection and their proliferative and differentiation potential. In fact, AMCs demonstrated high and rapid differentiation ability compared to MSCs. In addition, equine amnion-derived cells can also be frozen and recovered without loss of their functional integrity in terms of morphology and differentiation ability.

The noninvasive nature, low cost of collection, and the rapid proliferation along with a greater differentiation potential could make AMCs useful for cell therapy.

Comparative studies that employed cryopreserved heterologous AMCs and fresh autologous BM-MSCs in spontaneous equine tendon diseases in vivo showed that AMCs were well tolerated by horses, and all the clinical findings support the exertion of beneficial effects by the injected cells. Moreover, the tendon and ligament architecture is similar to that after injecting autologous BM-MSCs.

The encouraging results obtained with AMCs may result from the short interval between extraction and injection at the selected site, before any ultrastructural change occurs within the injured tendon. In fact, BM-MSCs when used as fresh autologous cells require prolonged in vitro culture, limiting the timeframe for implantation. The regenerated tissue could be less elastic and therefore functionally inferior to a normal or native tendon.

Speedy implantation, together with higher plasticity and proliferative capacity of the amniotic-derived cells compared to BM-MSCs, represent the main features of interest for this novel approach to the treatment of equine tendon diseases.

Moreover, for equine embryonic stem cells and fetal-derived equine embryonic-like stem cells, it is possible to speculate that AMCs, being less committed compared to adult cells, exhibit higher levels of engraftment with respect to BM-MSCs.

3.5 What Are the Limits?

Unfortunately, there are many limits to using large animal models in preclinical studies.

For example, there is limited availability of species-specific reagents, such as antibodies. Time-consuming experiments are usually needed to assess the reliability of results obtained using normally available reagents, because a very small number of antibodies are raised specifically against the porcine, bovine, and sheep antigen.

Growth factors and nutrients to be added in the culture media are rarely prepared and formulated for large animal species, and their efficacy in the latter must be tested. A further limit is the incomplete understanding of the networks controlling pluripotency and differentiation in large animals.

All these factors (see Fig. 3.9) make the task of designing a widely accepted protocol for routine applications very complex.

Another large problem is the lack of bona fide embryonic stem cells in domestic species. This limit has dramatically slowed the progression of experiments aimed at regenerative therapies that use large animals as a model. The many reasons at the base of the significant problems in generating stem cells in large animal species are discussed in other sections of this volume.

This concern has been obviated in part by the creation of induced pluripotent stem cells (iPSCs) from these species by standard reprogramming technologies

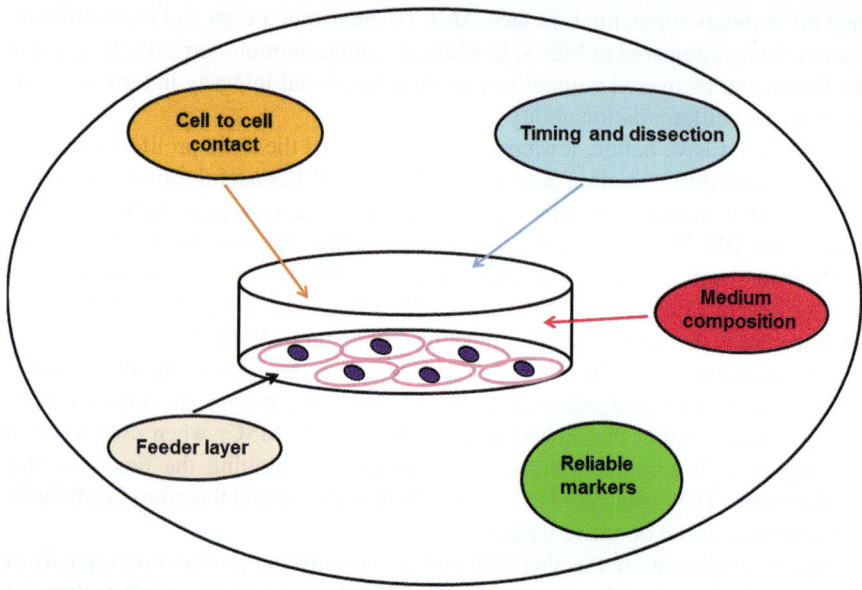

Fig. 3.9 Several issues must be addressed to obtain pluripotent cells in large animals: correct timing of isolation, and identification of standard methods for derivation, maintenance, and characterization of cell lines

Table 3.3 Stem cells derived in domestic species

	Putative embryonic stem cells (ESC)	Induced pluripotent stem cells (iPSC)	Adult stem cell
Ovine (sheep)	Dattena et al. (2006)	Bao et al. (2011)	Heidari et al. (2013)
Porcine (pig)	Brevini et al. (2007)	Wu et al. (2009)	West et al. (2010)
Bovine (cattle)	Pashaiasl et al. (2010)	Han et al. (2011)	Raoufi et al. (2011)
Canine (dog)	Hayes et al. (2008)	Luo et al. (2011)	Choi et al. (2013)
Equine (horse)	Saito et al. (2006)	Li et al. (2006)	Fortier and Smith (2008)

(Table 3.3). However, all these drawbacks have negatively impacted the wide and reproducible use of domestic species as intermediate experimental models.

The lack of centralized resources where cell lines obtained from large animal species can be characterized and stored represents a further limitation. Indeed, availability of reliable and specific reagents could help in creating consensus protocols and standardized procedures that would support the use of these alternative models. Maintenance of databases for the wider biomedical community would also help overcome these barriers and ensure an appropriate choice of animal models for particular human disease conditions and medical applications.

In an attempt to solve these problems, a group of international scientists recently founded the Domestic Animal Biomedical Embryology (DABE) Committee, which has the aims of serving as an informational resource for studies of embryological and developmental biology performed in domestic species and of working toward developing biomedical models for preclinical and translational research. DABE focuses on research addressed to the derivation, characterization, and differentiation of stem cells from farm and companion animals to provide a model for the development of cell-based therapies, the use of transgenesis and cloning for the generation of innovative biomedical models, and other research with similar scope.

The committee also provides a forum in which to exchange information on resources, reagents, and protocols that may prove more advantageous when applied to large animal species and that, as we have already discussed, represent a large concern. Interestingly, strong emphasis is also given to a clear definition of the international regulatory requirements for the safe and legal transport of tissues and cell lines for research purposes, which represent an area still greatly debated.

It is clear that a little step forward has been taken. However, more effort is needed and desirable to utilize this large range of animal models that would complement the mouse, allowing for a more comprehensive approach which could then be applied to humans.

Further Reading

Bao L, He L, Chen J, Wu Z, Liao J, Rao L, et al. Reprogramming of ovine adult fibroblasts to pluripotency via drug-inducible expression of defined factors. Cell Res. 2011;21(4):600–8.

Brevini T, Pennarossa G, Maffei S, Gandolfi F. Pluripotency network in porcine embryos and derived cell lines. Reprod Domest Anim. 2012;47 suppl 4:86–91. doi:10.1111/j.1439-0531. 2012.02060.x.

Brevini TA, Antonini S, Cillo F, Crestan M, Gandolfi F. Porcine embryonic stem cells: facts, challenges and hopes. Theriogenology. 2007;68 Suppl 1:S206–13.

Choi SA, Choi HS, Kim KJ, Lee DS, Lee JH, Park JY, et al. Isolation of canine mesenchymal stem cells from amniotic fluid and differentiation into hepatocyte-like cells. In Vitro Cell Dev Biol Anim. 2013;49(1):42–51.

Cibelli J, Emborg ME, Prockop DJ, Roberts M, Schatten G, Rao M, Harding J, Mirochnitchenko O. Strategies for improving animal models for regenerative medicine. Cell Stem Cell. 2013;12(3):271–4.

DABE: http://www.iets.org/comm_dabe.asp?autotry=true&ULnotkn=true

Dattena M, Chessa B, Lacerenza D, Accardo C, Pilichi S, et al. Isolation, culture, and characterization of embryonic cell lines from vitrified sheep blastocysts. Mol Reprod Dev. 2006;73:31–9.

Dixon JA, Spinale FG. Large animal models of heart failure: a critical link in the translation of basic science to clinical practice. Circ Heart Fail. 2009;2(3):262–71.

Doevendans PA, Daemen MJ, de Muinck ED, Smits JF. Cardiovascular phenotyping in mice. Cardiovasc Res. 1998;39(1):34–49.

Fortier LA, Smith RKW. Regenerative medicine for tendinous and ligamentous injuries of sport horses. Vet Clin North Am Equine Pract. 2008;24:191–201.

Gandolfi F, Vanelli A, Pennarossa G, Rahaman M, Acocella F, Brevini TA. Large animal models for cardiac stem cell therapies. Theriogenology. 2011;75(8):1416–25.

Gandolfi F, Pennarossa G, Maffei S, Brevini T. Why is it so difficult to derive pluripotent stem cells in domestic ungulates? Reprod Domest Anim. 2012;47 suppl 5:11–7. doi:10.1111/j. 1439-0531.2012.02106.x.

Han X, Han J, Ding F, Cao S, Lim SS, Dai Y, et al. Generation of induced pluripotent stem cells from bovine embryonic fibroblast cells. Cell Res. 2011;21(10):1509–12.

Harding J, Roberts RM, Mirochnitchenko O. Large animal models for stem cell therapy. Stem Cell Res Ther. 2013;4(2):23.

Hayes B, Fagerlie SR, Ramakrishnan A, Baran S, Harkey M, Graf L, et al. Derivation, characterization, and in vitro differentiation of canine embryonic stem cells. Stem Cells. 2008;26:465–73.

Heidari B, Shirazi A, Akhondi MM, Hassanpour H, Behzadi B, Naderi MM, et al. Comparison of proliferative and multilineage differentiation potential of sheep mesenchymal stem cells derived from bone marrow, liver, and adipose tissue. Avicenna J Med Biotechnol. 2013;5(2):104–17.

Kuzmuk KN, Schook LB. Pigs as a model for biomedical sciences. In: Rothschild MF, Ruvinsky A, editors. The genetics of the pig. 2nd ed. Wallingford: CAB International; 2011. p. 426–44.

Lange-Consiglio A, Corradetti B, Bizzaro D, et al. Characterization and potential applications of progenitor-like cells isolated from horse amniotic membrane. J Tissue Eng Regen Med. 2012;6:622–35. doi:10.1002/term.465.

Lange-Consiglio A, Corradetti B, Meucci A, et al. Characteristics of equine mesenchymal stem cells derived from amnion and bone marrow: in vitro proliferative and multilineage potential assessment. Equine Vet J. 2013. doi:10.1111/evj.12052.

Li X, Zhou SG, Imreh MP, Ährlund-Richter L, Allen WR. Horse embryonic stem cell lines from the proliferation of inner cell mass cells. Stem Cells Dev. 2006;15(4):523–31.

Luo J, Suhr ST, Chang EA, Wang K, Ross PJ, Nelson LL, et al. Generation of leukemia inhibitory factor and basic fibroblast growth factor-dependent induced pluripotent stem cells from canine adult somatic cells. Stem Cells Dev. 2011;20(10):1669–78.

McCall FC, Telukuntla KS, Karantalis V, Suncion VY, Heldman AW, Mushtaq M, Williams AR, Hare JM. Myocardial infarction and intramyocardial injection models in swine. Nat Protoc. 2012;7(8):1479–96.

Pashaiasl M, Khodadadi K, Holland MK, Verma PJ. The efficient generation of cell lines from bovine parthenotes. Cell Reprogram. 2010;12(5):571–9.

Pittenger MF, Mackay AM, Beck SC, et al. Multilineage potential of adult human mesenchymal stem cells. Science. 1999;284:143–7.

Raoufi MF, Tajik P, Dehghan MM, Eini F, Barin A. Isolation and differentiation of mesenchymal stem cells from bovine umbilical cord blood. Reprod Domest Anim. 2011;46(1):95–9. doi:10.1111/j.1439-0531.2010.01594.x.

Saito S, Sawai K, Minamihashi A, Ugai H, Murata T, Yokoyama KK. Derivation, maintenance, and induction of the differentiation in vitro of equine embryonic stem cells. Methods Mol Biol. 2006;329:59–79.

Suzuki S, Iwamoto M, Saito Y, Fuchimoto D, Sembon S, Suzuki M, Mikawa S, Hashimoto M, Aoki Y, Najima Y, Takagi S, Suzuki N, Suzuki E, Kubo M, Mimuro J, Kashiwakura Y, Madoiwa S, Sakata Y, Perry AC, Ishikawa F, Onishi A. Il2rg gene-targeted severe combined immunodeficiency pigs. Cell Stem Cell. 2012;10(6):753–8.

Ulrichs K, Eber S, Schneiker B, Gahn S, Strauß A, Moskalenko V, Chodnevskaja I. Isolation of porcine pancreatic islets for xenotransplantation. Methods Mol Biol. 2012;885:213–32. doi:10.1007/978-1-61779-845-0_13.

Webster RA, Blaber SP, Herbert BR, Wilkins MR, Vesey G. The role of mesenchymal stem cells in veterinary therapeutics: a review. N Z Vet J. 2012;60(5):265–72.

West FD, Terlouw SL, Kwon DJ, Mumaw JL, Dhara SK, Hasneen K, et al. Porcine induced pluripotent stem cells produce chimeric offspring. Stem Cells Dev. 2010;19(8):1211–20.

White FC, Roth DM, Bloor CM. The pig as a model for myocardial ischemia and exercise. Lab Anim Sci. 1986;36(4):351–6.

Wu Z, Chen J, Ren J, Bao L, Liao J, Cui C, et al. Generation of pig induced pluripotent stem cells with a drug-inducible system. J Mol Cell Biol. 2009;1(1):46–54.

Acknowledgments

Many scientists have dedicated their time and made precious comments about the text and concepts presented in this brief. Georgia Pennarossa, University of Milan, was engaged in daily base discussions and encouraged us with her help, criticism, and appreciation. Sara Maffei and Francesco Magri Bonvini provided great help in the preparation of images.

Special thanks to Cecilia Gandolfi for critical discussion and text editing. Thanks to the students of our Biomedical Sciences and Veterinary Medicine Courses. Their passion, enthusiasm, and curiosity supported the long hours spent at the microscope and under the sterile hood and enlightened even the heaviest days.

About the Authors

Tiziana A.L. Brevini is Associate Professor of Anatomy and Embryology at the University of Milan, Italy. She graduated in 1989 and spent 3 years at the Department of Molecular Embryology, Cambridge (UK). She obtained a Ph.D. in 1994 and then carried out research programs at Monash University, Melbourne and at the University of Adelaide, Australia. Her main area of research is addressed to the understanding of the cell differentiation process and pluripotency-related networks in mammalian cells and embryos.

T.A.L. Brevini and F. Gandolfi, *Pluripotency in Domestic Animal Cells*, SpringerBriefs in Stem Cells, DOI 10.1007/978-1-4899-8053-3, © Author 2013

Fulvio Gandolfi is Full Professor of Anatomy and Embryology at the University of Milan, Italy. He graduated in 1982 and spent several years at prestigious institutions in the UK, USA, and Australia. He is Editor in Chief of *Theriogenology*, Founding Member of the Inter-departmental Research Centre on Stem Cells (UniSTEM) of the University of Milano, Founder and Chairman of Domestic Animal Biomedical Embryology (DABE), and Vice-President of the International Congress of Animal Reproduction (ICAR). His main area of research focuses on early embryo development and cell commitment and differentiation.